Being Human
The Beginnings of Our Culture

The printing of this catalog was made possible by the Historisch-Archäologische Gesellschaft Frankfurt am Main e. V. Association for the Historisches- and Archäologisches Museum Frankfurt and the Research Center "The Role of Culture in Early Expansions of Humans" (ROCEEH) of the Heidelberg Academy of Sciences and Humanities

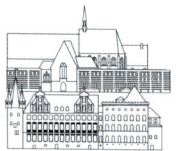

Historisch-Archäologische Gesellschaft
Frankfurt am Main e.V.

Archäologisches Museum Frankfurt – Publikationen 6
A publication of the Archäologisches Museum Frankfurt
and at the same time ROCEEH Perspectives 3
edited by Wolfgang David on behalf of the

148 pages with 136 illustrations and 16 icons

Bibliographic Information of the German National Library
The German National Library lists this publication in the
German National Bibliography; detailed bibliographic data
are available on the Internet at http://dnb.d-nb.de.

© 2022 by Nünnerich-Asmus Verlag & Media, Oppenheim am Rhein
ISBN 978-3-96176-193-7

Printed in Germany by Nünnerich-Asmus Verlag & Media GmbH

You can find more titles from our publishing program at:
www.na-verlag.de

Being Human

The Beginnings of Our Culture

Accompanying volume to the special exhibition

5 May 2021 – 27 March 2022
Archäologisches Museum Frankfurt

edited by Liane Giemsch and
Miriam Noël Haidle

**ARCHÆOLOGISCHES
MUSEUM** FRANKFURT

Content

Preface

Hardly a week goes by in the media without reports about new, sensational insights into the history of human development. Due to the rarity of new human fossil finds, these are usually reassessments of known finds as a result of new dating methods or molecular genetic analyses. Most of these reports relate to human biology and evolutionary lineage. The findings of prehistoric archeology, the sources of which are much more numerous than fossil human finds, and which also represent direct evidence of the culture of the oldest representatives of humankind, are rarely reported. However, early human history cannot be written solely on the basis of DNA or physical characteristics of the genus *Homo*. Because, from the beginning, it is also the story of our empowered mind and the culture we created in the all-encompassing sense, including social behavior, technologies, way of life, and nutrition, as well as aesthetic forms of expression. Thanks to geology, paleontology, paleoanthropology, and archeology, we now know that life existed on Earth for many hundreds of millions of years before our species developed in Africa. But what was the beginning of human culture, what provided the first impetus to this development in the course of which humans left behind their biological origins, left their original habitat, settled the entire world, and finally, in 1969, set foot on the moon? It is impressive to observe the constant acceleration of development, characterized by the ever-increasing human intervention into their environment and its transformation in so far as that we have now coined the term Anthropocene to distinguish a new geological era.

After Liane Giemsch, curator for prehistoric archeology, successfully organized the special exhibition "Gold and Wine. Georgia's Oldest Treasures" in 2018/19, I suggested that as a next project she should tackle a topic from her personal area of research, the older Paleolithic, to highlight and bring together her professional specialization, which is rare in Germany, and her rousing enthusiasm for the archeology of the oldest cultures in Africa, where she has carried out research in the past, in an exhibition. To my great delight, she suggested the earliest beginnings of our culture as a topic and immediately set up a student workgroup at the Goethe University in Frankfurt am Main for the project. Together with Miriam Noël Haidle, coordinator of the long-term project "The Role of Culture in Early Expansion of Humans" at the Senckenberg Research Institute and Natural History Museum in Frankfurt, and in cooperation with other well-known scientists, Liane Giemsch compiled the contents of the special exhibition and this companion volume. In the future, the topics highlighted here will be part of the new conception of the permanent exhibition at the Archaeological Museum in Frankfurt and its expansion into the digital space, to create a public forum at the interface between science and society about what it means to be human and the origins of human culture. Few fields of research are as dynamic as this one, as groundbreaking archeological discoveries continue to be made.

It began with the research of Jacques Boucher de Perthes (1788–1868), who in 1828 discovered handaxes, which he recognized as stone tools made by "antediluvian humans", together with the bones of extinct animals in the gravels of the Somme near Abbeville in the Picardy. The skeletal remains discovered in 1856 in the Neander Valley near Düsseldorf are thousands of years younger and, after a lengthy research dispute, were recognized as the remains of prehistoric humans. The skull of a pre-human of the species *Australopithecus africanus,* known as the "Taung child", was first discovered in South Africa in 1924. In 1931, Mary and Louis Leakey discovered the first evidence of a simple stone tool industry that is up to 2.6 million years old in the Olduvai Gorge in Tanzania. Just 30 years ago, in 1991, researchers discovered the 1.8 million-year-old fossils of *Homo erectus* and associated stone tools in Dmanisi, Georgia, which represented the oldest evidence of the genus *Homo* outside of Africa up until the more recent discovery of the 2.1 million-year-old stone tools at Shangchen in central China in 2018. And it was only in 2011, in Kenya at the Lomekwi site, that the worldwide oldest known stone artifacts, with an age of 3.3 million years, were found. These are significantly older than the oldest evidence of the genus *Homo* from around 2.8 million years ago.

In the first act of the influential science fiction film "2001: A Space Odyssey" by Stanley Kubrick (1928–99), entitled "The Dawn of Man", which was released in movie theaters in 1968, a few months before the first orbit of the moon by the Apollo 8 astronauts, a group of pre-humans in the African savanna came across a 4-million-year-old square-shaped black monolith that led to a decisive shift in their leader's consciousness. The exhibition "Being Human—The Beginnings of Our Culture" is on display in a former monastery from the late Middle Ages. This is a reminder that the questions raised in this exhibition are also aimed at philosophy and theology.

Dr. Wolfgang David M.A.
Director, Archäologisches Museum Frankfurt

Preface

Since humans became self-aware, they have asked fundamental questions: who are we and where do we come from? Time and again, the search for answers to these questions has led to fields of religion or philosophy, and these experiences and theories certainly play an important role for many people to this day when we ask, "what makes humans so special". Since the first fossil human finds were discovered in the middle of the 19th century, e.g., in the Neander Valley, and Charles Darwin and Russell Wallace developed the theory of evolution, these questions were also asked of natural sciences. Year after year, prehistoric archeologists and paleoanthropologists, together with colleagues from other disciplines, unearth new empirical data on the origins of humans and what makes us human. The exhibition *Being Human — The Beginnings of Our Culture* presents the most cutting-edge research on these questions and will therefore be of widespread interest.

The exhibition curators Dr. Liane Giemsch and PD Dr. Miriam Haidle are experts on these topics. Dr. Giemsch received her doctorate from the University of Tübingen with a thesis on the Paleolithic finds at Lake Manyara in northern Tanzania; her research focus is on the development of stone processing technologies. Dr. Haidle completed her habilitation at the University of Tübingen as a prehistoric scholar and paleoanthropologist with comparative studies on human and animal tool behavior and possible conclusions on the respective cognitive abilities. Both are closely linked to the long-term research project "The Role of Culture in Early Expansion of Humans" (ROCEEH) at the Heidelberg Academy of Sciences and Humanities based in Frankfurt and Tübingen. Dr. Giemsch was a member of the ROCEEH junior research group and is now curator for prehistoric archeology at the Archaeological Museum in Frankfurt. Dr. Haidle has coordinated the project from the Frankfurt ROCEEH office at the Senckenberg Research Institute since 2008, where she researches the development of cultural capacities. This exhibition, which is dedicated to being human and the beginnings of our culture, summarizes the results of the long-term work of the two curators and the ROCEEH team on the earliest phase of human development between 3.3 and 1 million years ago. It is the first of a series of syntheses, already planned, to present the last three million years of human history to a broad audience.

The first of these compilations, *Being Human*, starts at the beginning of human cultural history with the 3.3 million-year-old stone tools from Lomekwi in the Turkana region in Kenya. It illuminates the earliest phase of the Paleolithic and questions the origins of humans and their culture. The exhibition shows how culture shaped different areas of life from its inception: the skills and knowledge required for different technologies, the cooperation within a group, the inter-action with the environment, the use of resources. It emphasizes the role of learned behavior and the transmission of knowledge as main features of human development with roots that reach back to the oldest members of the genus *Homo* around three million years ago. *Being Human* discusses the importance of social learning and the gradual expansion of available actions based on the cumulative knowledge of many thousands of generations.

At this point, in addition to Dr. Giemsch and Dr. Haidle, I would also like to thank my colleagues in the ROCEEH team, apl. Prof. Dr. Michael Bolus, PD Dr. Angela Bruch, Dr. Christine Hertler, Julia Hess, Dr. Andrew Kandel, Prof. Dr. Friedemann Schrenk, and Dr. Christian Sommer, our ROCEEH guest PD Dr. Oliver Schlaudt, the director of the Archaeological Museum Frankfurt, Dr. Wolfgang David and his staff, the many students at the Goethe University in Frankfurt and numerous fellow scientists for their support and contribution to this exhibition and the supplementary volume about our origins, our identity, and our future.

Nicholas J. Conard
Speaker for the heads of research at ROCEEH
University of Tübingen & Senckenberg Center for Human Evolution
and Paleoenvironment

Liane Giemsch and Miriam Noël Haidle

In search of the beginnings of our culture

How did we become the humans we are today? When and where can we grasp the beginnings of our human existence for the first time? To find answers, let us shed light on the development of humankind in Africa between 3.3 and 1 million years ago. Two important physical changes had already developed in human-like species: an upright, bipedal locomotion and the resulting free hands. The construction of the hand, with short fingers and opposable thumbs, was ideally suited for handling various materials, objects, and tools. Today's great apes (chimpanzees, bonobos, orangutans, and gorillas) are also adept at manipulating their environment; that is probably part of our primate inheritance. Another part of this legacy is a long childhood with parents, aunts, uncles, family: an ideal breeding ground for social learning and culture. In the period between 3.3 and 1 million years ago, humans developed new ways to access their environment based on this heritage, which has had a lasting impact on their development to this day. Looking at this long history, it becomes clear how many different changes over the course of millions of years have contributed to making us the culturally diverse species that populate the entire planet today.

In addition to fossil skeletal remains, tools primarily shape our view of human developmental history. We often do not know which hominin form was responsible for individual archeological remains and thus for indications of certain behavior. Therefore, the following description of the developmental changes only provides a rough outline, without assigning the individual phases to specific hominin forms.

1 Human ancestral gallery. From top left to bottom right:
Sahelanthropus tchadensis, Australopithecus anamensis, Kenyanthropus platyops, Australopithecus afarensis, Australopithecus africanus, Paranthropus boisei, Homo rudolfensis, Homo habilis, Homo erectus, Neanderthal.

The first indications of the use of tools that is beyond the evidence known from great apes, date to the time a little over three million years ago. Tools were used to process stones at the Lomekwi site on Lake Turkana in Kenya. The resulting sharp edges made it easier to detach and cut up many things. The relationships with other species changed because of the resulting opportunities to procure plant and animal food. For example, hominins were now able to better compete with predators for parts of their prey. The proportion of animal food—in addition to meat, also the high-energy marrow from broken bones—increased. By using tools for various purposes, humans soon were able to assume a new unique position among carnivores as well as herbivores. They became real omnivores that could easily adapt depending on the situation.

Over time, we find evidence of an increased technical understanding both in the selection of raw materials and in the controlled processing of tools. Around 2.3 million years ago, hominins in Kenya were able to deliberately knap several dozen flakes from a single stone core. In the so-called Oldowan, they literally had the difficult-to-work material in hand. With these new capabilities as part of their skillset, humans first set foot in areas outside of Africa more than two million years ago. Around 1.8 million years ago, humans developed a new form of stone processing. In addition to manufacturing sharp-edged stone fragments, they started to rework the blanks into increasingly symmetrically and flat heavy-duty tools. Various new types of tools were created that were just as suitable for cutting up animal carcasses as for processing wood and other plant materials. Handaxes and other Acheulean tools played an important role for over 1.4 million years. And during this time humans also developed an ever-closer relationship with fire, the increasing use of which once again greatly changed their relationship with their environment.

To manufacture and use stone tools (or any other tools), humans had to learn many skills and acquire new knowledge: which raw material was suitable and where could it be found, what properties did a good hammerstone possess, how does one prepare a core to knap off a specific flake, which tool is best used for what purpose. None of these things were invented by one individual on their own. In a group, existing tools could be tried out for different tasks, existing methods were adopted and activities participated in, experienced users could set an example for others or intervene to help when needed. More complex actions became conceivable by breaking them down into modules. The special role of humans among the animals of the African savannah was shaped by their diverse and flexible integration into this environment as well as by their intensive social behavior, linked to an increased ability to learn. Growing group sizes and a closer community expanded the opportunities for learning in the social environment. Social togetherness became more diverse. The elders were not only role models in their actions but began to motivate, reinforce, and correct

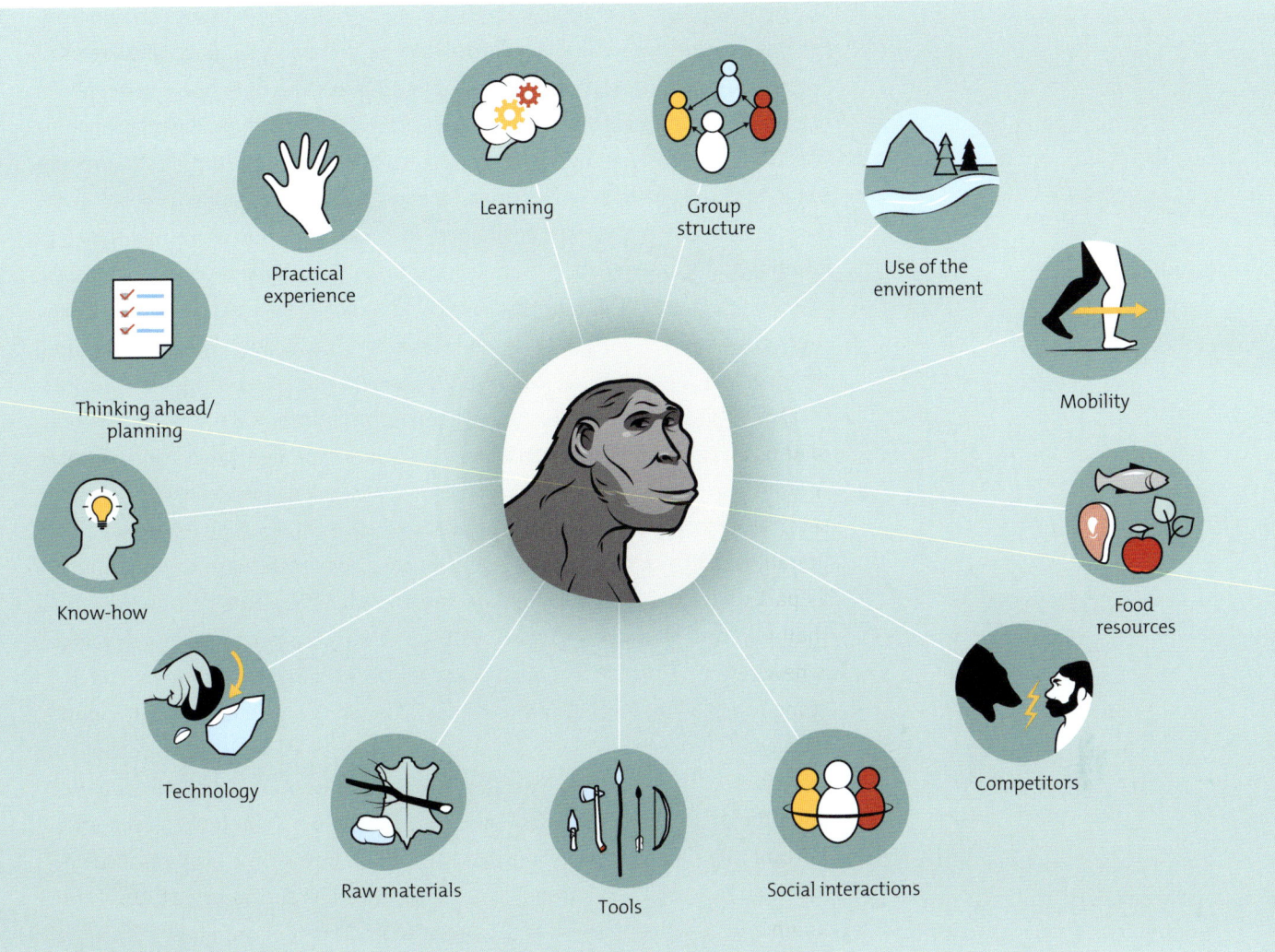

Practical experience

Learning

Group structure

Use of the environment

Mobility

Thinking ahead/ planning

Know-how

Food resources

Technology

Competitors

Raw materials

Tools

Social interactions

the inexperienced. In this way, more difficult things could be learned, and lengthy learning processes could be endured. Increasing social interaction required and enabled broader communication through gestures, facial expressions, and vocalizations. In a process that has lasted millions of years, the human ability to speak emerged slowly and with it the gift of not only passively demonstrating learning paths but also actively guiding them. Humans became cooperating partners, storytellers, and teachers.

But where is the culture? Is culture the combination of traditions and ideas? Is it a certain way of combining and maintaining skills and knowledge? Or does culture already begin with simple processes that are repeated in the group? Can we already speak of culture when monkeys manufacture and use simple tools to fish for termites or crack nuts? Where is the boundary between habit, tradition, and culture, or does it not exist at all? Cultural skills play an important

2 An excerpt of the universe of early human culture. The recovered stone tools and bones from prey animals only scratch the surface. Many different factors as well as social and material development processes were involved in dealing with the environment.

role in the formation of the genus *Homo* and the development of their environmental relationships. We can assume that all prehistoric and early humans of the last 3.5 million years possessed cultural abilities to different extents which manifested themselves in different forms of behavior.

Culture begins well before art, music, religion, and philosophy. It is not the product of an action that is endowed with special properties, but the doing itself. Cultural behavior (performance) is characterized by the development within the social environment and its relative durability. Culture is a socially learned practice in dealing with oneself, with one another, and with the environment that is communicated over generations. It is made up of many individual performances — actions and habitus. Culture is not something aloof but permeates everyday life. Humans have been cultural beings for millions of years, and they have evolved through and with culture.

Let us now look back at the beginnings of human culture, as much as we can grasp today, to discover some of the early key points on our developmental path.

Further Reading

Haidle, M. N./Hertler, C. 2021 KULT-UR-MENSCH. Kulturkonzepte für die Erforschung der Menschwerdung. ROCEEH Communications 1 (Heidelberg 2021).

Hörning, K. H./Reuter, J. 2004 Doing Culture: Kultur als Praxis. In: K. H. Hörning/J. Reuter (Hg.), Doing Culture. Neue Positionen zum Verhältnis von Kultur und sozialer Praxis. (Bielefeld 2004) 9 – 15.

Sahelanthropus tchadensis

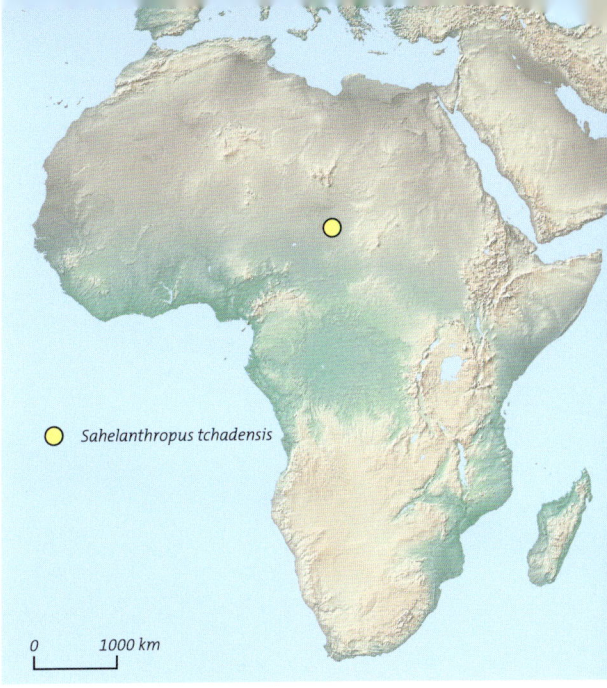

Discovery

In 2001, Ahounta Djimdoumalbaye discovered an almost completely preserved, yet heavily fragmented skull in the Djurab Desert (Chad).

Sites

Chad: Djurab Desert, Toros Menalla.

Finds

Skull without the lower jaw bow, four lower jaw bone fragments, and four isolated teeth. The fossils are all heavily damaged.

Age

circa 7 million years.

Brain size

360–370 cm³.

Characteristics

Sahelanthropus tchadensis probably lived in grass and forest landscapes and presumably ate mainly leaves, roots, and tubers. It is probable that they also ate large insects and small vertebrates when food was scarce. It is unclear whether the species was already permanently bipedal.

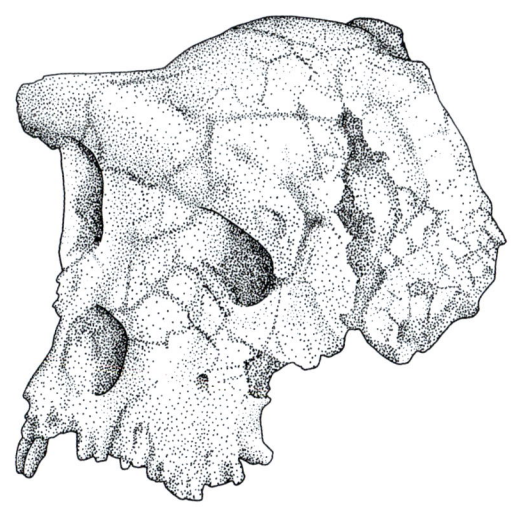

Skull from Toros-Menalla, Djurab Desert, Chad

Facial reconstruction

Fig:1.

Thomas Junker

Between nature and culture: the two origins of humanity

Man in the rudest state
in which he now exists
is the most dominant animal
that has ever appeared on this earth.
Charles Darwin, The Descent of Man (1871)

If you look at the human body — the external appearance, the anatomical details, and the physiological mechanisms — it is easy to see that humans are part of the animal kingdom. More precisely: they are mammals and primates. But that's not all. If one observes the behavior and way of life, then it is just as obvious that humans have strayed from their biological origins in many ways.

This peculiar amalgamation of nature and culture, which is so characteristic of humans, is fascinating and puzzling at the same time. And it can explain to a certain extent why it took so long for the biological origin of humans to be generally accepted. And this although the naturalists of the 18th century had already observed that the human body corresponds to that of other mammals and, above all, to that of primates in every detail.

Fig. 2

One species among many

For the founder of biological systematics, Carl Linnæus, these similarities allowed only one conclusion: in the first edition of his *Systema Naturæ* (System of Nature) from 1735, he classified humans in the animal kingdom. The species *Homo sapiens*, as he called it, was assigned a first rank, but was placed among the four-legged animals. In later editions, Linnæus changed some of the classifications and introduced the term "mammals" which is commonly used today.

1 The first scientific study of a chimpanzee by the doctor Edward Tyson
was published in 1699.

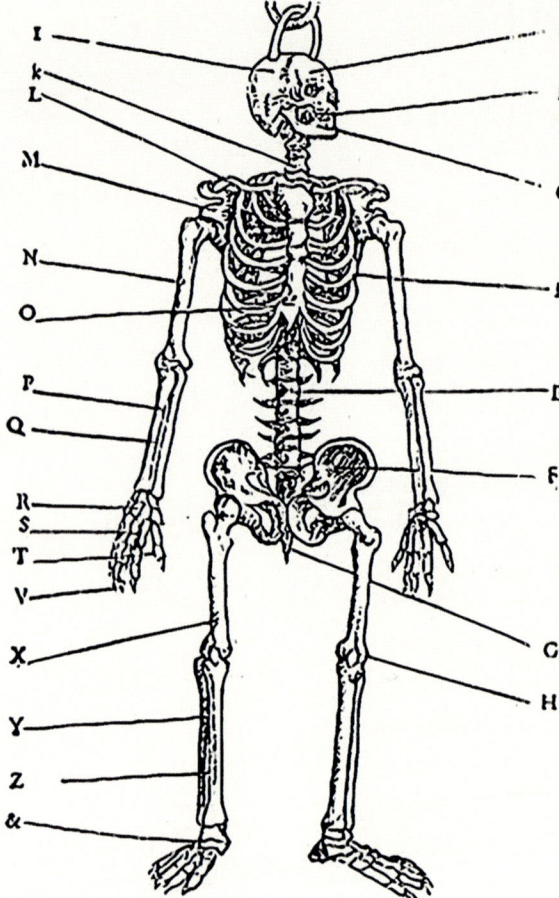

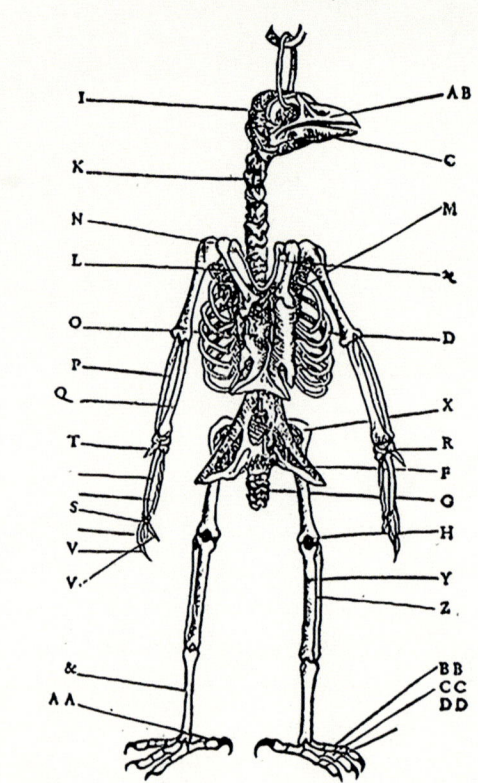

Portraict de l'amas des os humains, mis en comparaison de l'anatomie de ceux des oyseaux, faisant que les lettres d'icelle se raporteront à ceste cy, pour faire apparoistre combien l'affinité est grande des vns aux autres.

DES OYSEAVX, PAR P. BELON.

La comparaison du susdit portraict des os humains monstre combien cestuy cy qui est d'vn oyseau, en est prochain.

Portraict des os de l'oyseau.

A B Les Oyseaux n'ont dents ne leures, mais ont le bec tranchant fort ou foible, plus ou moins selon l'affaire qu'ils ont en à mettre en pieces ce dont ils viuent.

M Deux palerons longs & estroicts, vn en chascun costé.

Ↄ L'os qu'on nomme la Lunette ou Fourchette n'est trouué en aucun autre animal, hors mis en l'oyseau.

D Six costes, attachees au coffre de l'estomach par deuls, & aux six vertebres du dos par derriere.

F Les deux os des henches sont longs, caril n'y a aucunes vertebres au destoubs des costes.

G Six ostelets au cropion.

H La rouelle du genoil.

I Les sutures du test n'apparoissent gueres sinon qu'il soit bouilly.

k Douze vertebres au col, & six au dos.

d iii

2 Treatise by Pierre Belon from 1555. The similarities in the blueprint of organisms from different animal groups were observed at an early stage and today are important evidence of their common descent.

But regarding the point that had earned him the most criticism, he was not deterred: humans were part of the system of nature, and they were close to the apes. In many ways, the Linnæus system was still an uncertain first step. At the same time, however, it marks the beginning of an ideological revolution, the consequences of which were only slowly emerging in people's consciousness. From now on they were a part of nature, one species among many.

As a result, some scientists made it their life's work to find an absolute physical difference between humans and other animals—the number and arrangement of bones, the structure of the brain, or other properties—but each of these 'findings' turned out to be deceptive. What they found were quantitative deviations —in the proportions of arms and legs, in the hair and pigmentation of the skin, and the relative size of the brain. But no qualitative anatomical or physiological uniqueness.

The ape ancestry of humans

Linnæus did not explain the similarities between humans and other primates by their common evolutionary origin but believed that each species had been created separately. Some of his contemporaries were less hesitant, and soon people began to speculate about humans as modified apes and vice versa. The theory of evolution did not gain acceptance until a century later when Charles Darwin was able to show how the properties of living beings change in the interplay between heredity and selection. The natural system thus became the basis for the family tree of organisms.

It was only a small step from the conviction that humans are primates to the thesis that they descended from primates. Of course, they did not evolve from a primate species living today, but from a long line of primate ancestors that goes back more than 80 million years to the time of the dinosaurs. The exciting question was no longer whether, but from which fossil primate humans emerged. It was one of the great successes of molecular biology that, by comparing proteins and DNA, it was able to determine both the parentage and the approximate times of the separation. The now generally accepted result is that humans are most closely related to chimpanzees and that the last common ancestor lived five to seven million years ago.

Who were the last ape-like ancestors of the first humans? Who is the "ape" from whom we descended? Since the lineages of humans and chimpanzees separated around five to seven million years ago, but the first humans emerged around 2.5 million years ago, a gap of several million years remains. During this long time, our ancestors had already separated from the chimpanzees but were not yet humans. What were they then? Today they are described as an independent type of great ape, as australopithecines ("southern apes"). They were already able to walk upright, but there was no significant enlargement of the brain or other typical human features. The best-known representative of the species is "Lucy" (*Australopithecus afarensis*), who lived in East Africa 3.2 million years ago. Our last still ape-like ancestors were the australopithecines.

Fig. 3

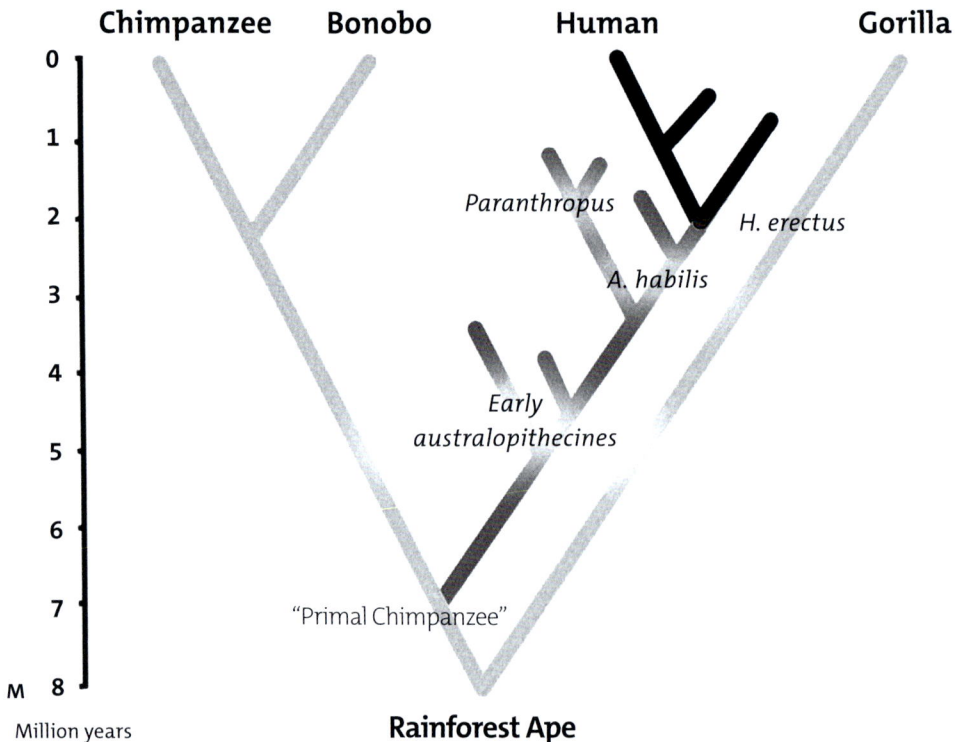

3 Simplified family tree of the African great apes.

Light line:
Chimpanzees and gorillas.

Dark line:
Australopithecines.

Black line:
Humans *(genus Homo).*

Chimpanzee   Bonobo   Human   Gorilla

Paranthropus

H. erectus

A. habilis

Early australopithecines

"Primal Chimpanzee"

Rainforest Ape

Million years

Limits of biology?

If one accepts that humans are primates and descend from ape-like ancestors, then the riddle of what it means to be human is not yet solved—on the contrary. Because then the question arises as to how and why this particular animal species, humans, with its unique characteristics, developed. Theologians and philosophers emphasize to this day that biology is not in a position to solve this riddle. It is argued that a leap occurred in the course of evolutionary development which cannot be explained scientifically, and which resulted in an absolute difference between humans and other animals. In the Catholic Church, for example, the natural evolution of the human body is accepted while at the same time emphasizing that the spirit soul must have been created directly by God.

Absolute differences between humans and animals were also postulated in philosophy and other humanities. For the French philosopher and naturalist René Descartes, for example, the body of all living things was "a kind of machine", made up of bones, nerves, muscles, veins, blood, and skin. There is only one exception in this physically determined system: the indivisible and immortal human soul. Similar ideas are still alive today. The cultural philosopher Ernst Cassirer wrote that the "symbolic forms"—language, myth, science, religion, technology, art—are "true primordial phenomena of the spirit" that cannot be explained causally.

Fig. 4

4 The painting *"Pithecanthropus alalus"* by Gabriel von Max (1840–1915) from 1894 represents a fictional transitional form between the ape-like ancestors and today's humans.

Scientists and evolutionary biologists on the other hand tried to close the gap between humans and other animals. They argued that humans are shaped by their evolutionary heritage not only physically, but also in their feelings, thoughts, and behavior. It is basic biological knowledge that there is a close connection between the physical characteristics of an animal and its behavior. In principle, this applies to all areas of life, and humans are no exception.

The fact that humans have abilities that are only rudimentarily found in other animals—language, art, and science, for example—contradicts this only at first glance. From a biological point of view, humans have unique characteristics— just like all other living beings are special and unique in their own way. Never-

5 There are also traditional cultural behaviors in animals. One example is the method of termite fishing among chimpanzees.

theless, the question arises whether the extraordinary human characteristics can be explained by general evolutionary mechanisms or whether the method reaches its limits here. In the following, I would like to briefly discuss this using the example of culture and show to what extent the more recent biological theories build a bridge between the natural sciences and humanities approaches.

How much nature is in culture?

In terms of evolutionary biology, the cultural ability can be defined as an adaptation that combines the advantages of genetic information with those of individual experiences and at the same time avoids some of its disadvantages. What does that mean? In genetic inheritance, genes are the information-carrying units. They produce relatively inflexible behavior that can only be changed through mutations, recombination, and selection. In contrast, learned behaviors are more flexible. This can be beneficial when an animal lives in a changing environment. However, learned behavior has a serious disadvantage: each individual must re-live and learn from the experiences again and again. And this can be associated with great risks, e.g. when learning which food is edible and where there is danger.

Social animals can compensate for this disadvantage by learning from other group members and adopting their experiences. In this way, a second information storage is created, the units of which are not inherited genetically, but conveyed through example and upbringing—culture. Therefore, cultural ability can be defined as a social learning ability and as such is genetically determined, an adaptation.

Fig. 5

Individual experiences and cultural knowledge are important additions to the evolutionary knowledge fixed in the genes. However, this means that the learned behavior and the thoughts associated with it must not be genetically determined but must be free and open to new and unexpected things. Only then can they meaningfully complement the genetically determined instincts and behaviors.

What happens when the learned behavior interferes with biological functions? Then the individual will hurt itself or die. The same applies to cultures. The Shakers, a Christian free church that flourished in the USA in the early 19th century and had several thousand members, are an example of this. The basis of their coexistence was celibacy and complete sexual abstinence. Whatever one thinks this way of life morally, it is not biologically sustainable. In general, this means that if a society lives by cultural beliefs and rules that conflict with biological necessities, that culture will sooner or later die out.

So humans live in two worlds: they are both natural and cultural beings. In this respect, biologists are correct when they point out that human culture arose naturally and will disappear again when it no longer fulfills its purpose. But the philosophers and humanities scholars are also right when they point out that the cultural content, the special thoughts, and convictions, are not genetically determined. In this sense, humans and other animals that learn from experience can think and behave freely. However, there are always risks associated with freedom. And so, the two origins of humans from nature and culture are an ongoing evolutionary experiment with an unknown outcome: for each individual, for nations and their cultures, for humans as a biological species.

Further reading

Belon, P. 1555 L'histoire de la nature des oyseaux (Paris 1555).

Bonner, J. T. 1980 The evolution of culture in animals (Princeton 1980).

Cassirer, E. 1925 Sprache und Mythos.—Ein Beitrag zum Problem der Götternamen. In: E. Cassirer, Wesen und Wirkung des Symbolbegriffs (Darmstadt 1956) 71–158.

Conard, N. J. (Hg.) 2006 Woher kommt der Mensch? 2. Aufl. (Tübingen 2006).

Darwin, C. 1871 The descent of man, and selection in relation to sex. 2 vols. (London 1871).

Descartes, R. 1641 Meditations on first philosophy (Copenhagen 2020).

Haeckel, E. 1889. The history of creation: or the development of the earth and its inhabitants by the action of natural causes (London 2018).

Junker, T. 2018 Die Evolution des Menschen. 3. Aufl. (München 2018).

Junker, T./Hoßfeld, U. 2009 Die Entdeckung der Evolution: eine revolutionäre Theorie und ihre Geschichte. 2. Aufl. (Darmstadt 2009).

Linnæus, C. 1735 Systema naturæ, sive regna tria naturæ systematice proposita per classes, ordines, genera & species (Leiden 1735).

Tyson, E. 1699 Orang-outang, sive *Homo sylvestris*: or, the anatomy of a pygmie compared with that of a monkey, an ape, and a man (London 1699).

Ardipithecus ramidus and *kadabba*

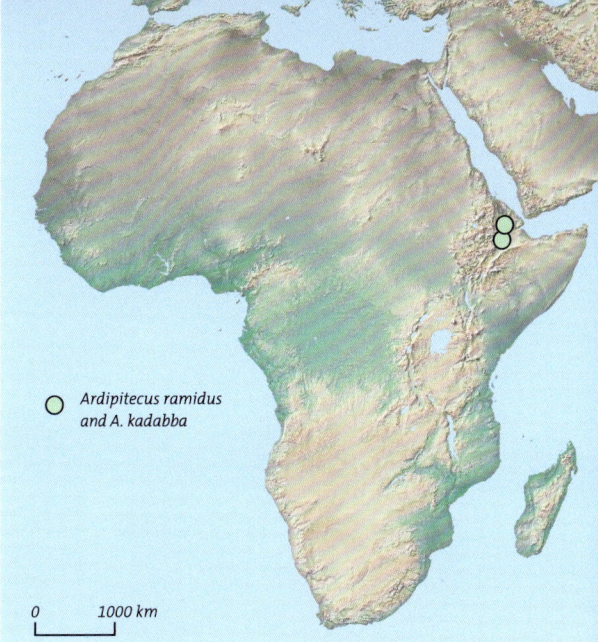

Ardipitecus ramidus and A. kadabba

0   1000 km

Discovery

In 1992, Gen Suwa discovered a first molar of this species in Aramis, Ethiopia. Another associate set of ten teeth was found in 1993. "Ardi"—a largely preserved skeleton—was discovered between 1994 and 1996 in the Afar Triangle in Ethiopia and represents a truly sensational find. Yohannes Haile-Selassie discovered the first fossil of an *Ardipithecus kadabba* in 1997 in the Afar Valley in Ethiopia.

Sites

Ethiopia: Aramis, Awash River

Finds

ramidus: Complete skeleton, teeth.

kadabba: Right lower jaw fragment with molar. Four additional isolated teeth from the lower jaw were discovered at a later point in time.

Age

ramidus: 4.42–3.9 million years.

kadabba: 5.8–5.18 million years.

Brain size

280–350 cm³.

Characteristics

Since there are up to 1.9 million years between the fossil records of *Ardipithecus kadabba* and *Ardipithecus ramidus,* it is assumed that they are two different species. *Ardipithecus* represents an early link between the climbing locomotion of the great apes and the constant bipedal walk of humans. The splayed toe and the con-struction of the pelvis show that they still retained their climbing ability despite walking on two legs. It is not entirely clear what they ate. The thickness of the enamel and the width of the upper incisors suggest that they ate less fruit than today's chimpanzees, but more ripe fruit, succulent plant parts, and young leaves than *Australopithecus afarensis.*

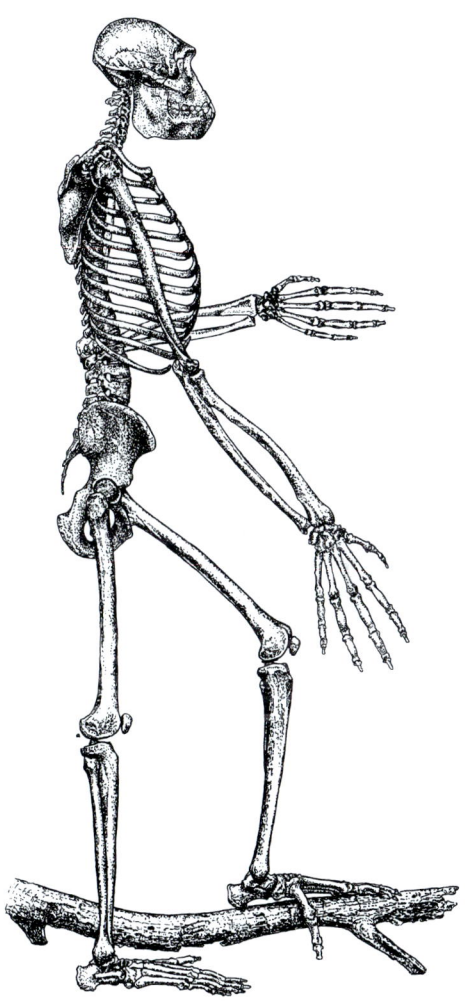

Reconstruction of *Ardipithecus ramidus*

Friedemann Schrenk

Early human biocultural evolution

350,000 generations of human history (approx. 7 million years) document a great geographical diversity of pre-human, early, and prehistoric humans, initially in Africa and beyond for circa the past 100,000 generations. Climate and environmental changes as well as changes in food resources were often the triggers for new developments.

Walking upright: the origin of hominins

Great apes lived in the rainforests of tropical Africa, which stretched from the west to the east coast of the continent, for over 30 million years. Climbing, but not brachiation, was one of the common traits of modern great apes. Our ancestors never lived "in the trees", but were a four-legged species who could straighten up and probably also stand for a short period of time.

Due to global climate cooling since the Middle Miocene around 10 Ma (= million years ago), the tropical rainforest shrank, so that some great ape populations newly colonized the emerging African savannas. The fruit-rich food of the tropical rainforest was partially replaced by aquatic food. Since great apes cannot swim, they waded into the shallow water to gather food, which in the long term further stabilized the two-legged locomotion. The shore habitats of the savannah were thus the ideal breeding ground for bipedal walking.

As the savannas extended across more than 5 million km^2 it is unlikely that the bipedal walk developed only once. All finds from this period (Kenya, *Orrorin*, approx. 6 Ma, Ethiopia, *Ardipithecus*, approx. 5.8 Ma, and Chad, *Sahelanthropus*, approx. 7 Ma) show evidence for bipedal walking. The geographical variants of *Fig. 2* the original populations of the earliest hominins were intertwined along the borders of the shrinking tropical rainforest. The reduction in canine teeth is also an early feature of hominin origin. This suggests a changed social behavior in which social cognition and higher forms of cooperation were able to develop.

1 To recover even the smallest bone fragments, archeologists sieve the sediment from a site in Malawi.

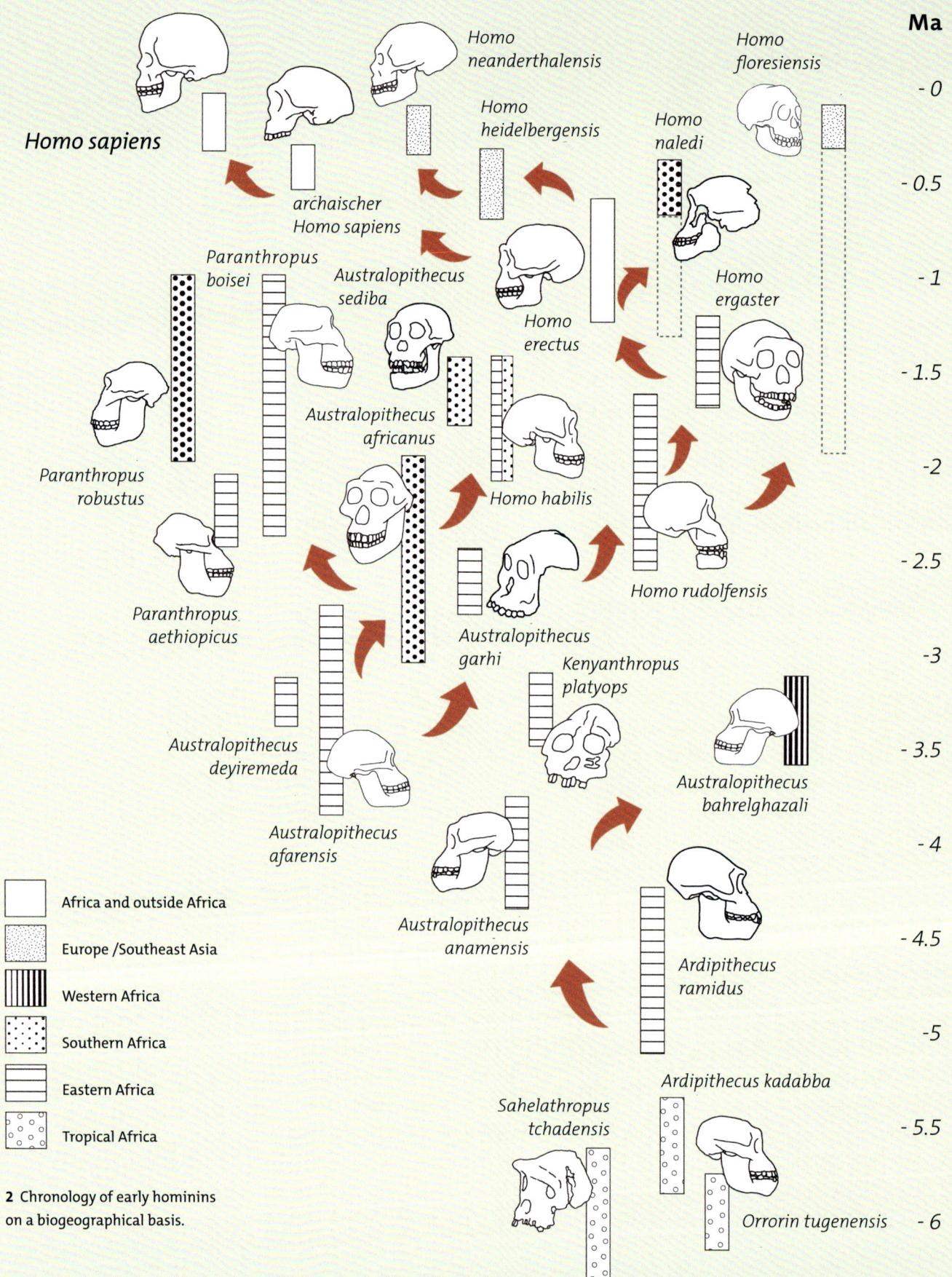

Homo sapiens

Homo neanderthalensis

Homo floresiensis

archaischer Homo sapiens

Homo heidelbergensis

Homo naledi

Paranthropus boisei

Australopithecus sediba

Homo erectus

Homo ergaster

Paranthropus robustus

Australopithecus africanus

Homo habilis

Paranthropus aethiopicus

Australopithecus garhi

Homo rudolfensis

Kenyanthropus platyops

Australopithecus deyiremeda

Australopithecus bahrelghazali

Australopithecus afarensis

Australopithecus anamensis

Ardipithecus ramidus

Sahelathropus tchadensis

Ardipithecus kadabba

Orrorin tugenensis

	Africa and outside Africa
	Europe /Southeast Asia
	Western Africa
	Southern Africa
	Eastern Africa
	Tropical Africa

2 Chronology of early hominins on a biogeographical basis.

Ma

- 0
- 0.5
- 1
- 1.5
- 2
- 2.5
- 3
- 3.5
- 4
- 4.5
- 5
- 5.5
- 6

Pre-humans in eastern and western Africa

The oldest pre-human finds of the genus *Australopithecus* were discovered on the southeastern shore of Lake Turkana in Kenya (*Australopithecus anamensis*, approx. 4 Ma). All australopithecines had a brain no larger than that of chimpanzees, large molars, and thick enamel. The teeth and jaws were suitable for chewing hard and brittle food or for crushing small particles, such as nuts and other seeds, between their flat, broad molars.

Since neither physical nor cultural achievements provided effective defense, a pronounced cooperative social behavior took over the decisive protective function against predators. Australopithecine finds from Laetoli (Tanzania) and Hadar (Ethiopia) were jointly used as the basis for the description of *Australopithecus afarensis* (3.7–2.9 Ma). *Australopithecus afarensis* (for example "Lucy") weighed 30 to 50 kg and was about 1.20 m tall. The arms were relatively long, the legs very short compared to those of modern humans. The fully developed upright locomotion was therefore still quite strenuous.

Fig. 3

The acquisition of food was probably relatively unspecialized: fruits, berries, nuts, seeds, saplings, buds, and mushrooms were available. Underground roots and tubers could be dug up. Small reptiles, fledglings, eggs, mollusks, insects, and small mammals living in the water and on the ground were also not spurned. Due to the seasonal change, *Australopithecus afarensis* is likely to have developed strategies to make the best possible use of the diverse food supply according to the availability in a seasonal habitat.

Fig. 4

The pre-humans gradually achieved pan-African distribution, but always remained close to the broad riverbank habitats. *Australopithecus deyiremeda*, *Australopithecus garhi*, and *Kenyanthropus platyops* originated in eastern Africa. A subpopulation expanded into modern-day Chad *(Australopithecus bahrelghazali).*

Pre-humans in southern Africa

In periods of relatively warm climates around three and a half to three million years ago, pre-human populations also spread along coastal corridors into southern Africa. The first hominin discovery in Africa ("Taung Baby", 1925) led to the first description of the genus *Australopithecus*. The mouth region protrudes, the face is tilted slightly (prognathic). The forehead is flat, the bulge above the eye developed. The lateral cheekbones project powerfully, the jaw is robust, the chin is missing. A characteristic trait is a combination of a small cranium (approx. 450 cm³) with a set of teeth in which the incisors and canines appear tiny, while the molars and premolars are almost twice as large as in modern humans.

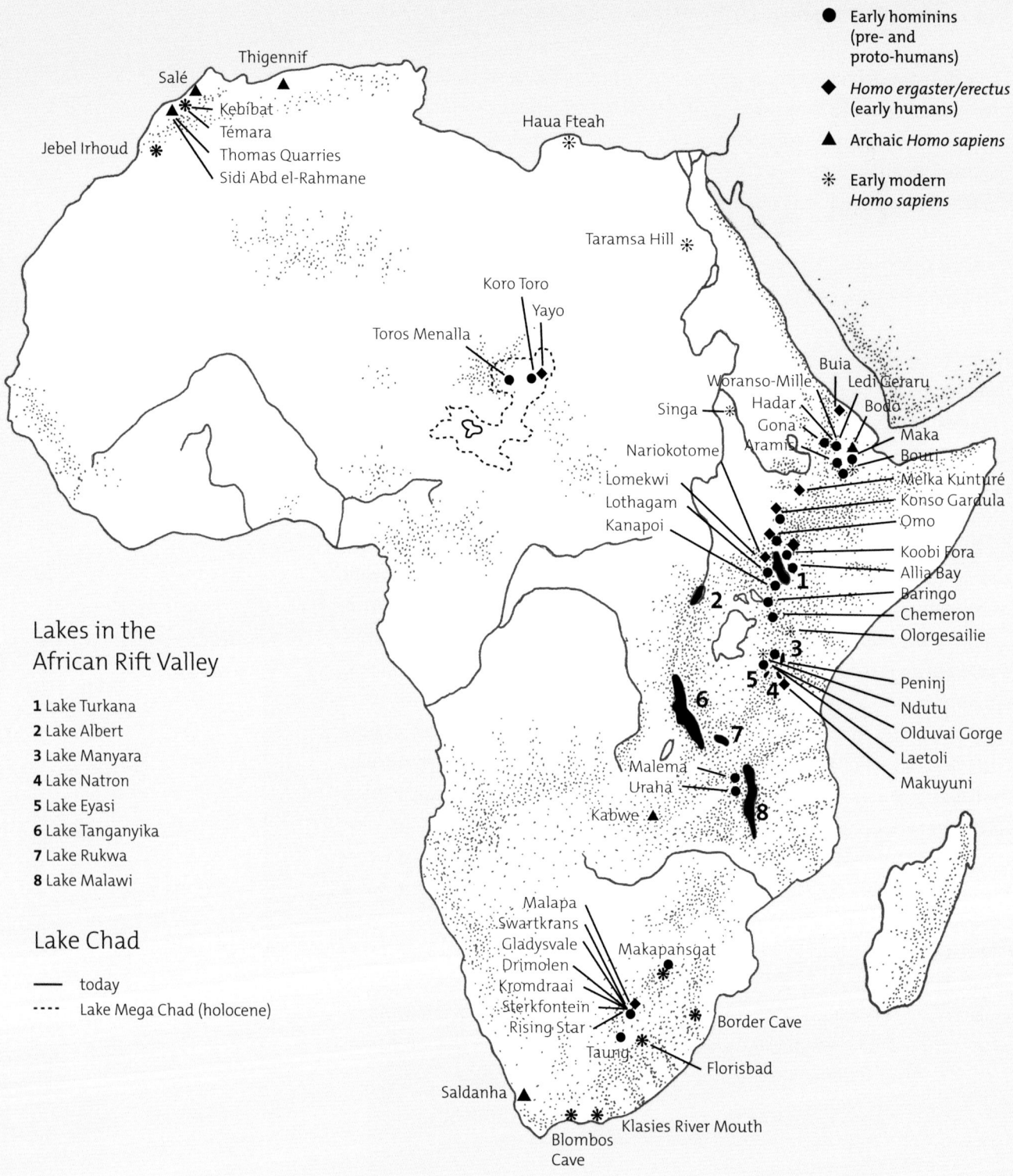

Thigennif

Salé

Kebibat
Témara
Thomas Quarries
Sidi Abd el-Rahmane

Jebel Irhoud

Haua Fteah

Taramsa Hill

Koro Toro
Yayo

Toros Menalla

Buia
Woranso-Mille   Ledi Geraru
Hadar   Bodo
Singa   Gona   Maka
Aramis   Bouri
Nariokotome   Melka Kunturé
Lomekwi   Konso Gardula
Lothagam   Omo
Kanapoi   Koobi Fora
Allia Bay
Baringo
Chemeron
Olorgesailie
Peninj
Ndutu
Olduvai Gorge
Laetoli
Makuyuni

Malema
Uraha
Kabwe

Malapa
Swartkrans
Gladysvale
Drimolen   Makapansgat
Kromdraai
Sterkfontein   Border Cave
Rising Star
Taung   Florisbad

Saldanha

Blombos
Cave   Klasies River Mouth

● Early hominins
(pre- and
proto-humans)

◆ *Homo ergaster/erectus*
(early humans)

▲ Archaic *Homo sapiens*

✳ Early modern
Homo sapiens

Lakes in the
African Rift Valley

1 Lake Turkana
2 Lake Albert
3 Lake Manyara
4 Lake Natron
5 Lake Eyasi
6 Lake Tanganyika
7 Lake Rukwa
8 Lake Malawi

Lake Chad

——— today
- - - - Lake Mega Chad (holocene)

Australopithecus africanus in southern Africa preferred habitats along the forest edge, often near rivers. There is no evidence for hunting behavior, therefore it is likely that only smaller animals or freshly torn game were eaten. Presumably, pre-humans drove away predators in a cooperative and targeted manner, for example by throwing stones. Pre-humans ate everything they could get their hands on in an opportunistic manner, with varying proportions of plants and meat according to the season.

Fig. 2

Climate change catalyst: expansion, evolution, culture

A phase of global cooling began around 2.8 million years ago. For around 15,000 generations, up to around 2.5 million years ago, pre-humans lived in increasingly extreme climatic and environmental conditions, which led to a profound change in the food base and a geographical shift in habitats. This resulted in passive expansions and evolutionary adaptations as well as the beginning of the biocultural evolution of the genus *Homo*.

Passive expansion: Some organisms retained their preference for seasonal changes by expanding towards the equator along with the shrinking biome (ecosystem). These "passive migrants" also included sub-populations of *Australopithecus africanus*, which spread north along corridors along the riverbanks. *Homo habilis* evolved due to its greater flexibility in behavior in the new living space.

Evolutionary adaptation: Some populations of *Australopithecus afarensis* in eastern Africa were able to digest the harder foodstuffs that were abundantly available in the open habitats, using their large molars. They developed wide facial bones and megadont (oversized) dentition. The zygomatic arches were strong and wide. The conspicuous formation of a sagittal crest on the top of the skull served as the attachment area for the greatly enlarged lateral masticatory muscles *(Musculus temporalis)*. Their megadont molars show that they chewed predominantly hard and coarse vegetable foods, such as seeds and hard plant fibers. The ability to break open hard shells could also have been beneficial when consuming aquatic food (such as mussels). The robust "nutcracker people" *aethiopicus, boisei,* and *robustus* are grouped in the genus *Paranthropus.*

Fig. 5

4 Comparison of early hominin skeletons.

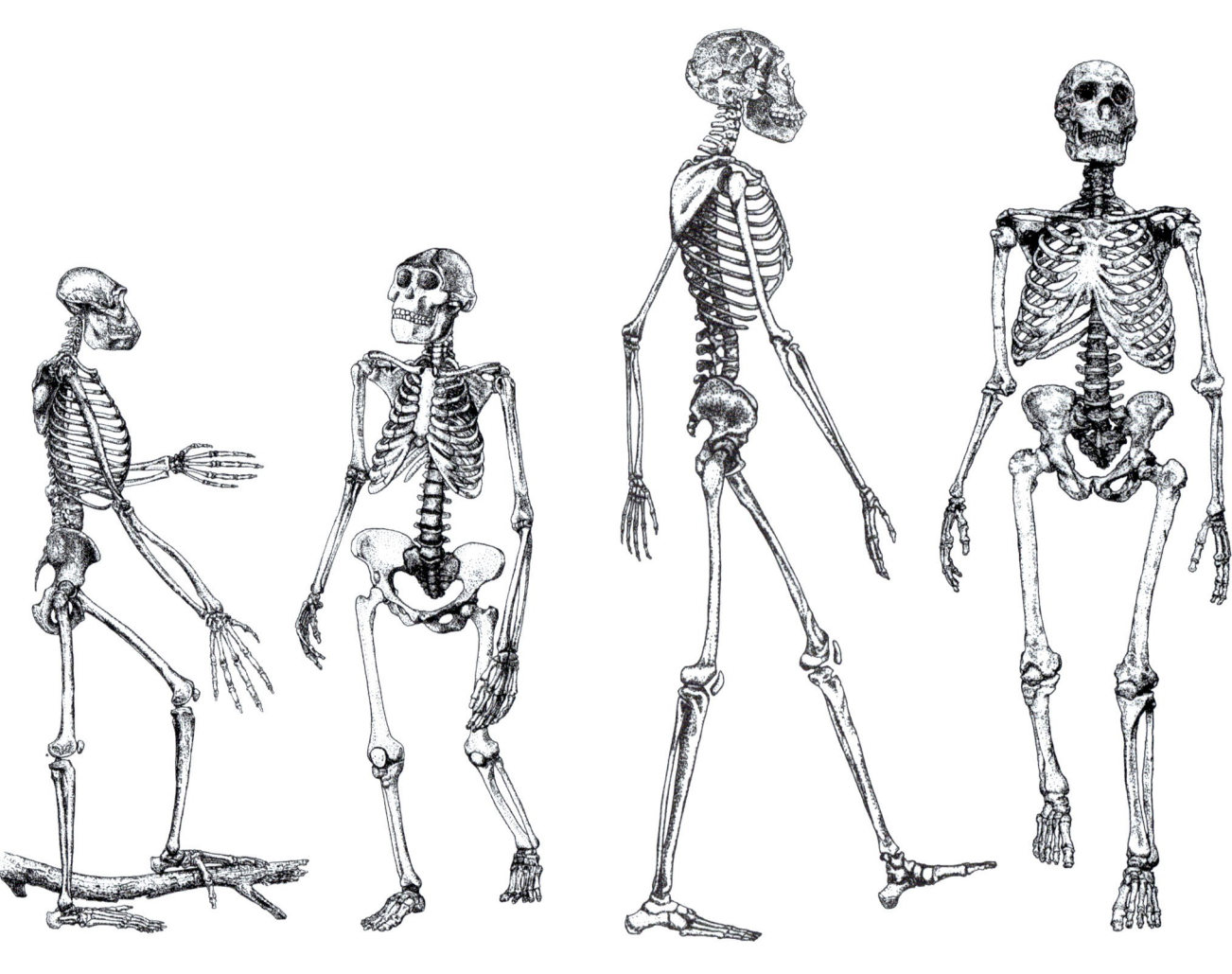

Reconstruction
Ardipithecus ramidus
Height: circa 1.20 m
Age: circa 4.4 million years

Reconstruction of "Lucy"
Australopithecus afarensis
Height: circa 1,20 m
Age: circa 2.9 million years

Early African
Homo erectus (Homo ergaster)
"Turkana Boy", skeleton KNM-WT 15000 from Nariokotome, West-Turkana, Kenya
Height: circa 1.70 m
Age: circa 1.7 million years

Homo neanderthalensis
Reconstruction of the skeleton, using La Ferrassie 1 (France) and Kebara 1 (Israel)
Height: circa 1.60 m
Age: circa 70,000–60,000 years

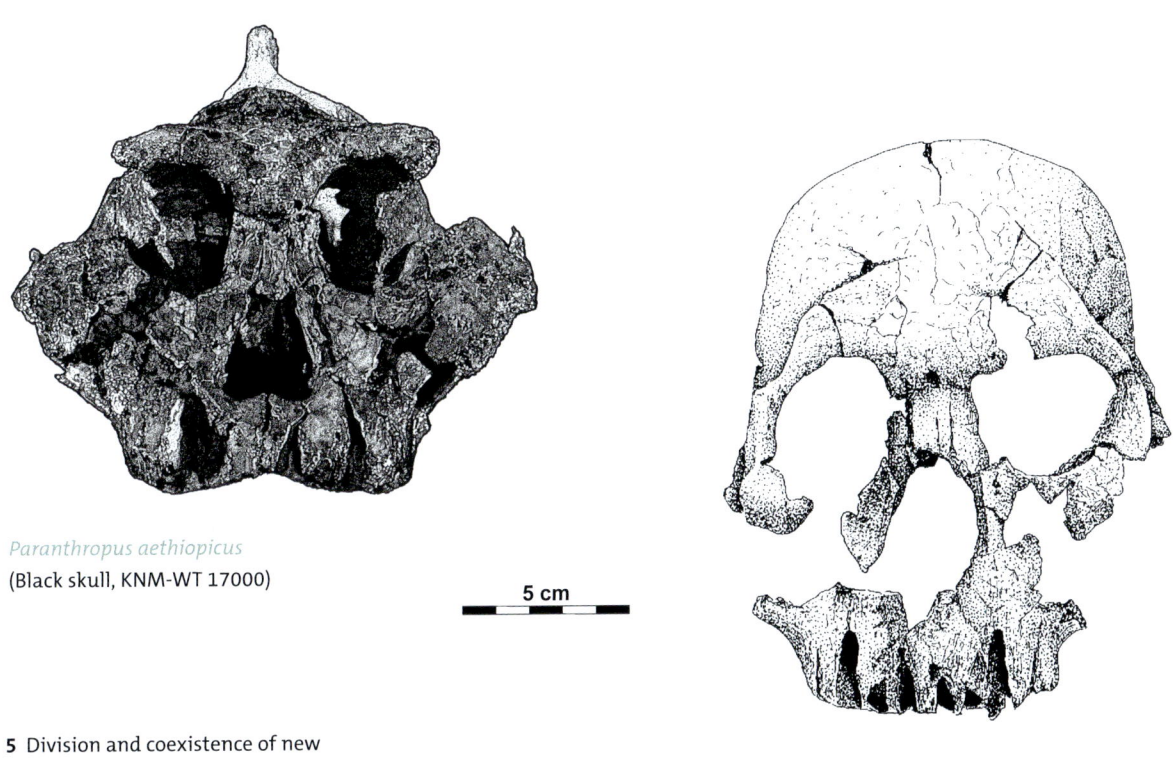

Paranthropus aethiopicus
(Black skull, KNM-WT 17000)

5 cm

Homo rudolfensis
(KNM-ER 1470)

5 Division and coexistence of new species in the hominin family tree for 2.5 million years.

Biocultural evolution: There was, however, an alternative to the hyper-robust chewing apparatus that was also suitable for chopping up increasingly harder food: the use of tools. Under the pressure of habitat changes over the past 2.8 million years, it was the hominins' capability of cultural behavior that gave birth to the genus *Homo*. The oldest prehistoric humans belong to the species *Homo rudolfensis*. By systematically using stones to crush the hard plant food, these prehistoric humans gained increasing independence from direct environmental influences. However, this inevitably led to a growing dependency on tools—a characteristic of humans to this day.

6 Expansions and important find sites of the genus *Homo* in Europe, Africa, and Asia.

Uraha: UR 501
Lower jaw
Homo rudolfensis
Age circa 2.5 million years
Malawi

W-Turkana: KNM-WT 15000
Skeleton
Homo ergaster
Age circa 1.7 million years
Kenya

E-Turkana: KNM-ER 1470
Skull
Homo rudolfensis
Age circa 2 million years
Kenya

Olduvai: OH 9
Skull cap
Homo erectus
Age circa 1 million years
Tanzania

Bodo: Bodo cranium
Skull
archaic *Homo sapiens*
Age circa 600,000 years
Ethiopia

Kabwe: Broken Hill 1
Skull
archaic *Homo sapiens* and
Homo rhodesiensis
Age circa 300,000 years
Zambia

Jebel Irhoud:
Virtual skull reconstruction
Homo sapiens
Age circa 300,000 years
Morocco

Qafzeh: Qafzeh IX
Skull
Homo sapiens
Age circa 95,000 years
Israel

EUROPE

Atapuerca: Gran Dolina
Skull fragment and Upper jaw
Homo antecessor
Age circa 800,000 years
Spain

Atapuerca: Sima de los Huesos
Skull 5
Homo heidelbergensis
Age circa 400,000 years
Spain

Mauer: Mauer 1
Lower jaw
Homo heidelbergensis
Age circa 600,000 years
Germany

Petralona: Skull
Homo erectus and *heidelbergensis*
Age circa 300,000 years
Greece

Arago: Arago 21
Skull
Homo erectus and *heidelbergensis*
Age circa 400,000 years
France

Steinheim:
Skull
Homo heidelbergensis and
steinheimensis
Age circa 350,000 years
Germany

Gibraltar: Gibraltar 1
Skull
Homo neanderthalensis
Age circa 25,000 years
Great Britain

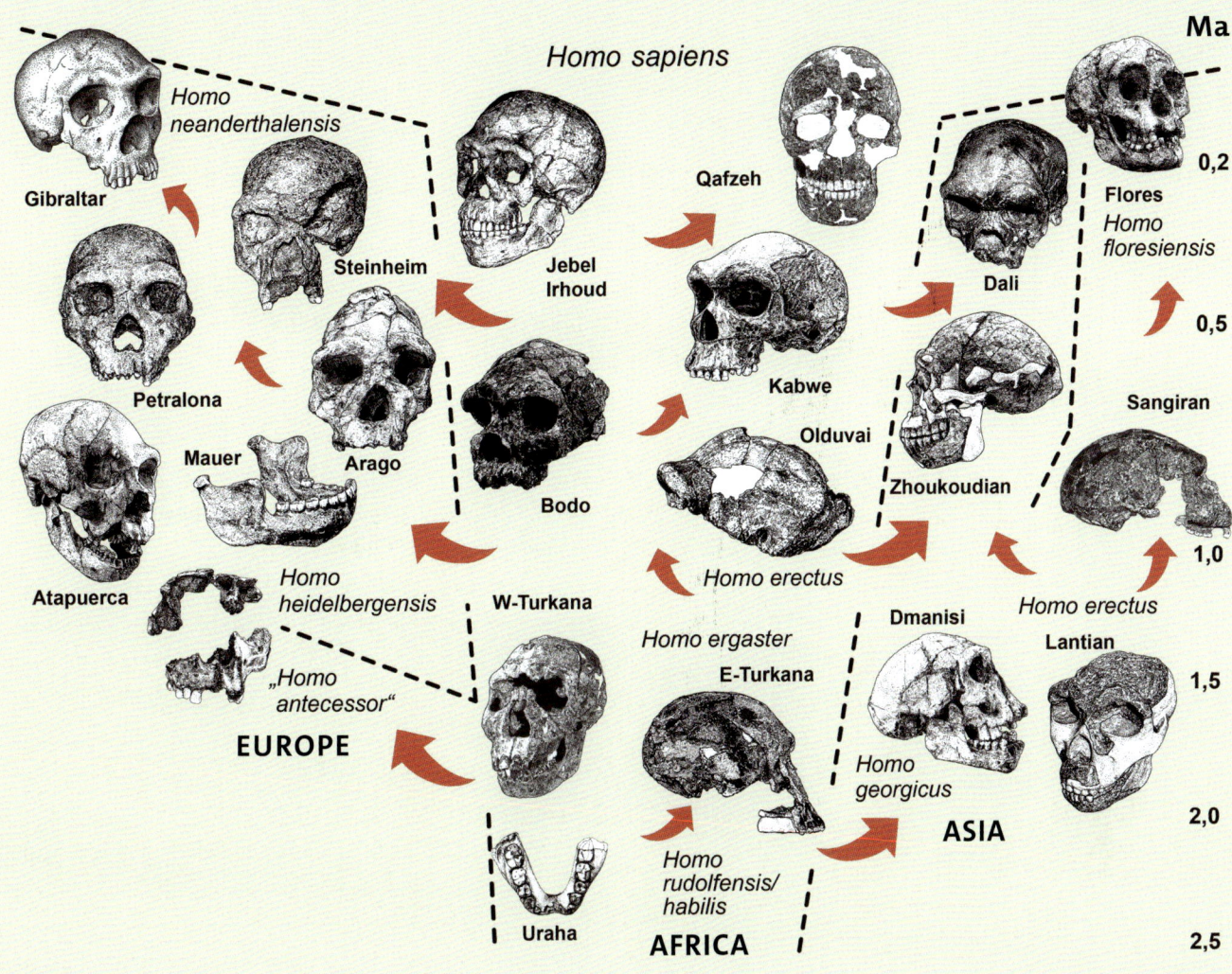

Homo sapiens

Ma

Gibraltar

Homo neanderthalensis

Steinheim

Qafzeh

Jebel Irhoud

Dali

Flores
Homo floresiensis

0,2

Petralona

Kabwe

0,5

Mauer   Arago

Olduvai

Zhoukoudian

Sangiran

Atapuerca

Bodo

Homo heidelbergensis   W-Turkana

„Homo antecessor"

EUROPE

Homo erectus

Homo erectus

1,0

Homo ergaster

Dmanisi

Lantian

E-Turkana

Homo georgicus

ASIA

1,5

Homo rudolfensis/ habilis

2,0

Uraha   **AFRICA**

2,5

ASIA

Lantian: Gongwangling
Skull
Reconstruction
Homo erectus
Age circa 2 million years
China

Dmanisi: D 2700 & D 2735
Skull and Lower jaw
Homo georgicus
Age circa 1.8 million years
Georgia

Sangiran: Sangiran 17
Skull
Homo erectus
Age not determinable
Indonesia

Zhkoudian:
Skull, Reconstruction
Sinanthropus pekinensis and
Homo erectus
Age circa 600,000 years
China

Dali: Dali skull
Skull
archaic *Homo sapiens*
Age circa 280,000 years
China

Flores: Liang Bua (LB) 1
Skull
Homo floresiensis
Age circa 50,000 years
Indonesia

Early humans: *Homo erectus*

Fig. 3

About two million years ago in Africa, the development towards a stronger and larger skeleton and massive bone structure, the typical traits of *Homo erectus*, began. The oldest fossil remains, between 2 and 1.5 million years old, were described as *Homo ergaster.* The volume of the skull gradually increased, and the proportions of the cranium and facial skeleton changed. The point where the spinal column and spinal cord connect to the skull *(Foramen magnum)* moved further underneath the skull, the structure of the joint of the lower jaw changed, and the more rounded shape of the dental arch emerged. The massive bone structure shows that *Homo erectus* displayed great strength and endurance when carrying material and food. These early humans could run, as evidenced by the elongation of legs; loss of body hair and increase in sweat glands probably also developed during the *Homo erectus* phase.

The finger bones are elongated and no longer suitable for climbing. In contrast to the great apes' "power grip", which allowed them to clasp an object with their fingers and thumb, a "precision grip" is now possible. Due to the shorter fingers and the greater flexibility of the thumb, this is now opposable in such a way that the fingertips can touch each other. With more control, it is now possible to precisely manipulate objects held in one hand.

The earliest finds possess a brain volume of around 800 to 900 cm^3. The volume increased to around 900 to 1,000 cm^3 circa one million years ago, and more than 1,100 to 1,200 cm^3 half a million years ago. The more efficient brain improved the ability to store and process complex connections. There are no direct references to language among *Homo erectus*. Given the ability to produce tools that require a great deal of experience and knowledge to make, it can be assumed that verbal communication also increased. The increased demand for energy required by a larger brain necessitates an omnivorous lifestyle with a high proportion of meat. One way to efficiently digest plant foods is through the application of fire. The earliest evidence for the controlled use of fire was discovered in Koobi Fora, Kenya (approx. 1.5 Ma). The control of the fire is a technical and at the same time socially and proactively regulated task (see article "Fire" by Giemsch in this volume). We can assume a well-functioning social structure for *Homo erectus*.

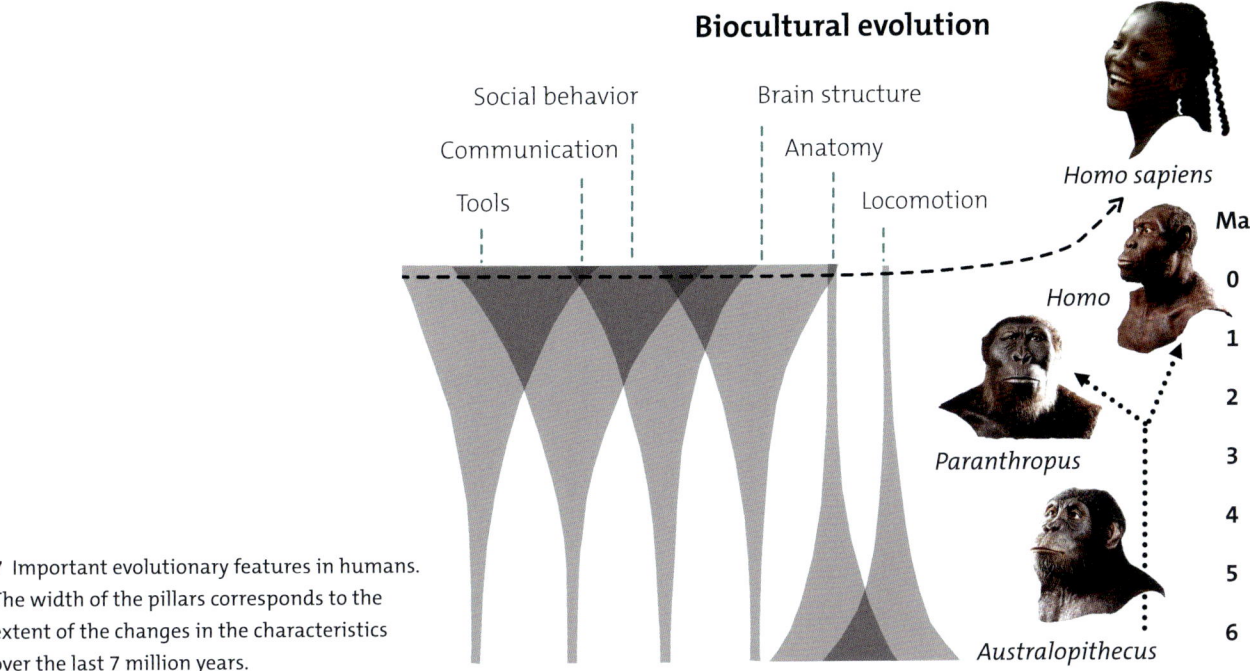

Biocultural evolution

Social behavior

Communication

Tools

Brain structure

Anatomy

Locomotion

Homo sapiens

Ma

Homo

0

1

Paranthropus

2

3

4

5

6

Australopithecus

7 Important evolutionary features in humans. The width of the pillars corresponds to the extent of the changes in the characteristics over the last 7 million years.

Earliest expansions 'Out of Africa'

Homo erectus was very familiar with the landscape and the availability of resources. This led to efficient use of seasonal resources and a greater radius of movement. Its ability to combine plant and meat resources gave it great flexibility long before the use of fire. The high tolerance towards variations in habitats led to the division into groups that lived further apart. River valleys allowed for a rapid spread. Expansions along the seacoast provided the opportunity to gather mollusks. Expansions of a few kilometers per generation led to the colonization of new habitats over a short geological period.

Fig. 6

Early humans left Africa for the first time around two million years ago, probably along routes through the Levant and the Arabian Peninsula. Evidence of the earliest settlements was found in China (2.1 Ma), Pakistan, and the Caucasus (Dmanisi, 1.8 Ma). Finds in Italy are 1.7 to 1.3 million years old, in Spain 1.4 to 1.2 million years. An increase of cultural abilities facilitated the expansion across habitat boundaries, such as into southeastern Asia (approx. 1.5 Ma) and the Philippines (approx. 700,000 years).

At least 500,000 years ago, *Homo erectus* was widespread in East and Southeast Asia as well as throughout Central and Southern Europe. While these geographical variants evolved into independent species (for example Neanderthals in Europe, Denisova people in Asia), in Africa around 400,000 years ago, a syn-

Fig. 7 ergy effect of different factors of biological and cultural evolution, such as tool culture, communication, social behavior, brain structure, and body composition, led to the emergence of modern humans. The transfer of knowledge and cultural and genetic exchange were the decisive prerequisites for innovations and the worldwide spread of *Homo sapiens*. Therefore today's isolation of affluent regions – thought in many generations – will not be successful. Given global challenges such as serious changes in biodiversity and climate, only global networking can ensure our survival, as our long history has shown time and again.

Further reading

Bonnefille, R. 2010 Cenozoic vegetation, climate change and hominid evolution in tropical Africa. Global and Planetary Change 72(4), 2010, 390–411. https://doi.org/10.1016/j.gloplacha.2010.01.015

Dean, C./Leakey, M.D./Reid, D./Schrenk, F./Schwartz, G.T./Stringer, C./Walker, A. 2001 Growth processes in teeth distinguish modern humans from *Homo erectus* and earlier hominins. Nature 414, 2001, 628–631. https://www.nature.com/articles/414628a

Ingicco, T./van den Bergh, G.D./Jago-on, S.C.B./Bahain, J.J./Chacón, M.G./Amano, N./Forestier, H. et al. 2018 Earliest known hominin activity in the Philippines by 709 thousand years ago. Nature 557, 2018, 233–237. https://doi.org/10.1038/s41586-018-0072-8

Lordkipanidze, D./Ponce de León, M.S./Margvelashvili, A./Rak, Y./Rightmire, G.P./Vekua, A./Zollikofer, C.P.E. 2013 A complete skull from Dmanisi, Georgia, and the evolutionary biology of early *Homo*. Science 342, 2013, 326–331. https://doi.org/10.1126/science.1238484

Lüdecke, T./Schrenk, F./Thiemeyer, H./Kullmer, O./Bromage, T.G./Sandrock, O./Fiebig., J./Mulch, A. 2016 Persistent C3 vegetation accompanied by Plio-Pleistocene hominin evolution in the Malawi Rift (Chiwondo Beds, Malawi). Journal of Human Evolution 90, 2016, 163–175. https://doi.org/10.1016/j.jhevol.2015.10.014

Maslin, M.A./Shultz, S./Trauth, M.H. 2015 A synthesis of the theories and concepts of early human evolution. Philosophical Transactions of the Royal Society B Biological Sciences 370, 2015, 2014–2064. https://doi.org/10.1098/rstb.2014.0064

Potts, R. 2013 Hominin evolution in settings of strong environmental variability. Quaternary Science Reviews 73, 2013, 1–13. https://doi.org/10.1016/j.quascirev.2013.04.003

Ring, U./Schrenk, F./Albrecht, C. 2018 The East African Rift System: tectonics, climate and biodiversity. In: Hoorn, C./Perrigo, A./Antonelli, A. (Hg.), Mountains, climate and biodiversity, (Chichester 2018) 391–411.

Schrenk, F. 2019 Die Frühzeit des Menschen, 5. Aufl., (München 2019).

Weston, E./Friday, A.E./Johnstone, R.A./Schrenk, F. 2004 Wide faces or large canines? The attractive versus the aggressive primate. Proceedings of the Royal Society of London B 271(6), 2004, 416–419. https://doi.org/10.1098/rsbl.2004.0203

Australopithecus afarensis

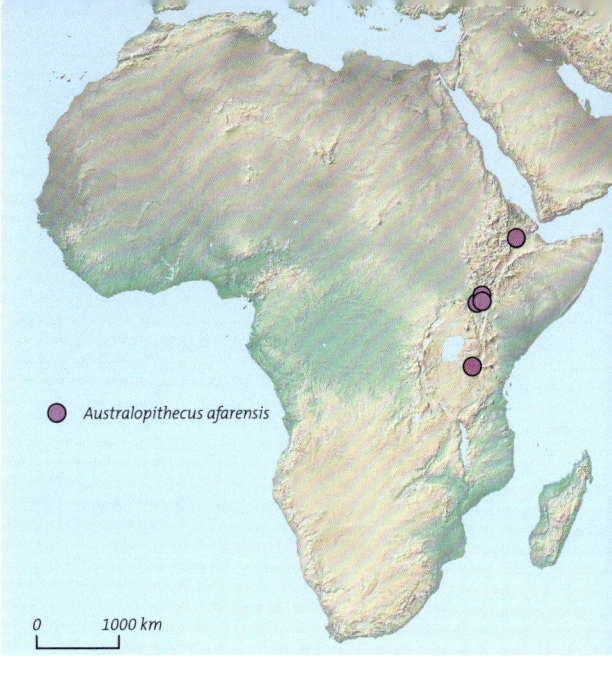

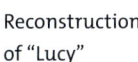
Australopithecus afarensis

0   1000 km

Discovery

Ludwig Kohl-Larsen first discovered remains, jaw fragments and teeth, in 1938/39 in the Garusi Valley near Laetoli, Tanzania. It was not until 1979 that Donald Johanson, the discoverer of Lucy, and colleagues described a separate species *Australopithecus afarensis* based on finds from Laetoli and Ethiopia.

Sites

Tanzania: Laetoli.

Ethiopia: Hadar, Maka, Dikika, Aramis, Mount Galili.

Kenya: Turkwel River.

Finds

Left lower canine, incomplete skeleton ("Lucy"), juvenile skeleton from Dikika, and additional hand, foot, and extremity fragments, as well as the fossilized footsteps from Laetoli.

Age

3.76–2.92 million years.

Brain size

450–550 cm³.

Characteristics

Australopithecus afarensis mainly lived in a so-called mosaic land-scape (grassland with isolated trees and closed stands of bushes and trees on waterways and mountain valleys). They reached a size of approximately 1.20–1.40 m and weighed 20–50 kg. This weight corresponds to that of today's dwarf chimpanzees. Their diet was based primarily on fruits, leaves, plant pulp, seeds, and herbs. The angle of the knee joint suggests that they walked upright. The anatomy of the finger and toe bones, which are shorter than those of monkeys, are indicative of life on the ground.

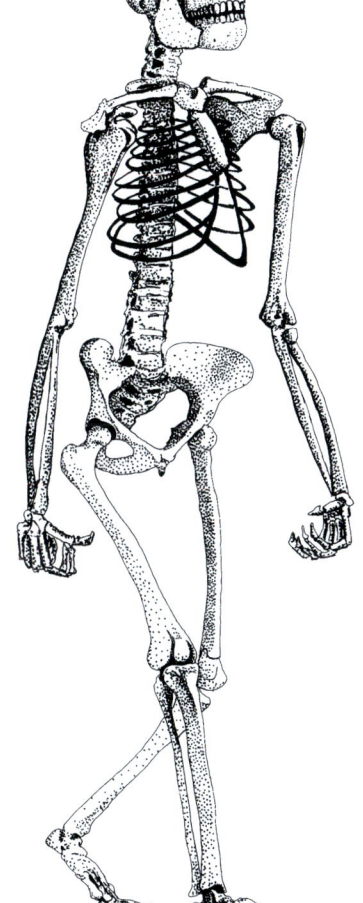

Reconstruction of "Lucy"

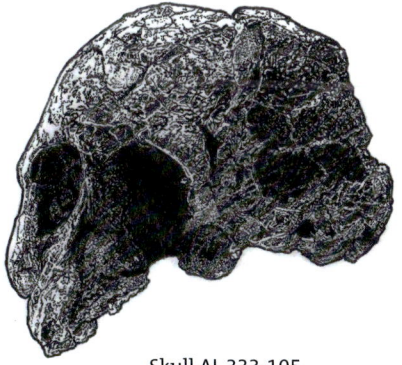

Skull AL 333-105
from Hadar, Ethiopia

Facial reconstruction

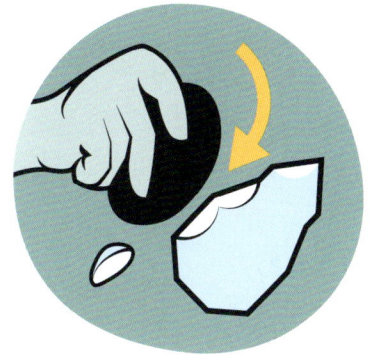

Michael Bolus

The earliest stage of human stone tool technology: the Oldowan

Introduction

The Oldowan belongs to the early part of the African Early Stone Age, the earliest development stage of human stone technology. It is named after the Olduvai (formerly also Oldoway) Gorge in Tanzania, whose history of prehistoric exploration is primarily associated with the couple Mary and Louis Leakey. In December 1931, L. Leakey presented the discovery of a simple stone industry from Bed I at Olduvai Gorge to the public for the first time, at that time using the term "pre-Chellean". He first coined the term Oldowan in 1936. The Leakeys' extensive research at Olduvai Gorge brought to light many additional find-bearing layers with artifacts from the Oldowan, and it was above all the research of Mary Leakey, especially in Bed I and Bed II of Olduvai, that provided us with an outline and a better understanding of the Oldowan period.

Until a few years ago, the Oldowan, with an age of around 2.6 million years, was considered to be the oldest-known stage of tool production, but recent excavations in Kenya have revealed even older artifacts. From around 1.8 million years ago onwards, the Oldowan existed in Africa parallel to the Acheulean, which is mainly characterized by tools such as bifacial handaxes that are worked on both surfaces, which are missing in the Oldowan (see the article "Acheulean" by Giemsch in this volume). Since worked pebbles are a significant part of the Oldowan assemblages, it is often referred to as a 'pebble tool' industry and as a Mode I industry based on the classification by Grahame Clark. We now know that the flakes knapped from these pebbles were at least as important as the pebbles. Against this background, the worked pebbles should be seen not so much as implement but as cores. Of course, this does not rule out that they were also used as crude tools.

Fig. 1

---

1 Different aspects of a chopping tool from Melka Kunture, Ethiopia.

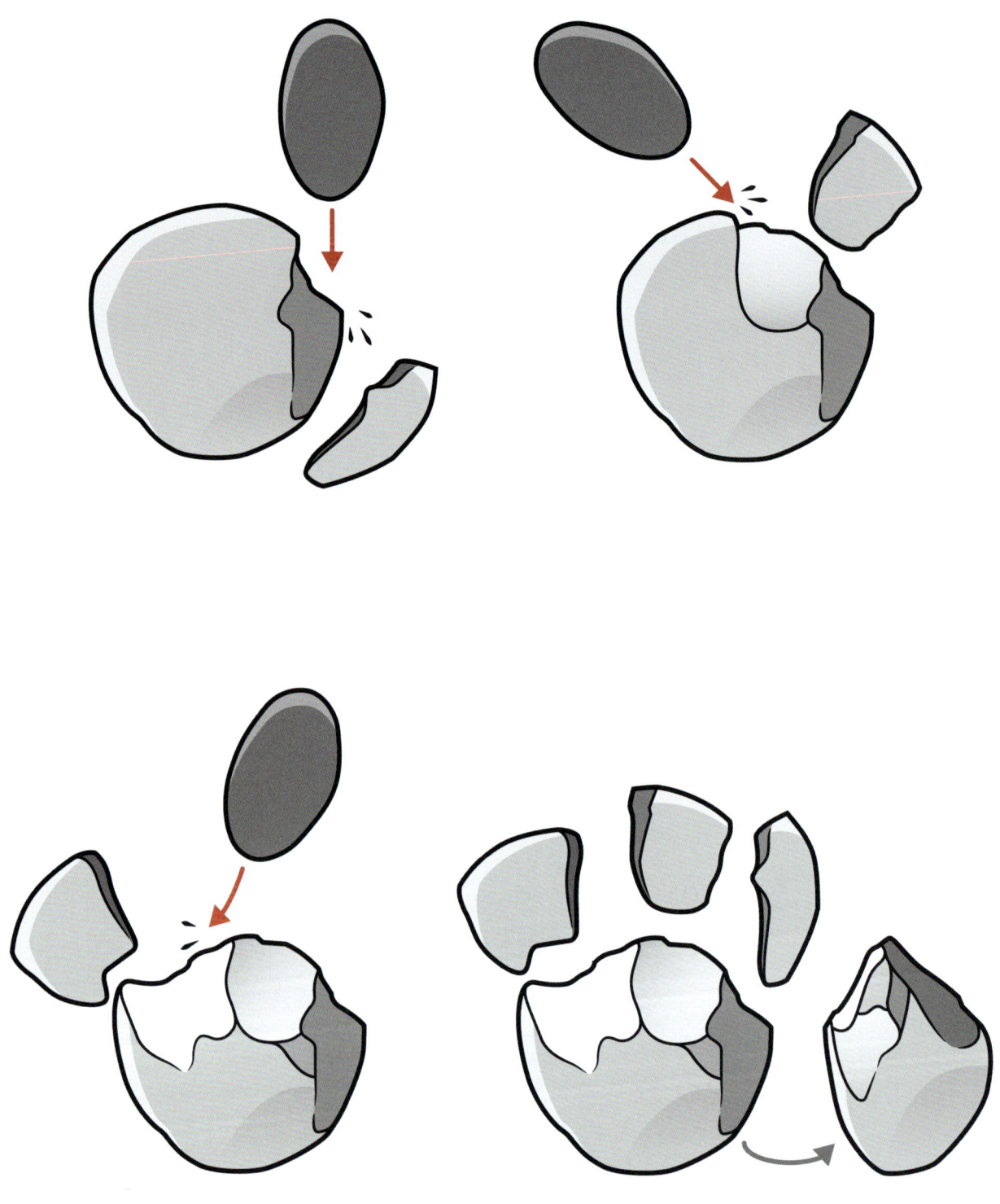

2 Production scheme of an Oldowan implement with flakes.

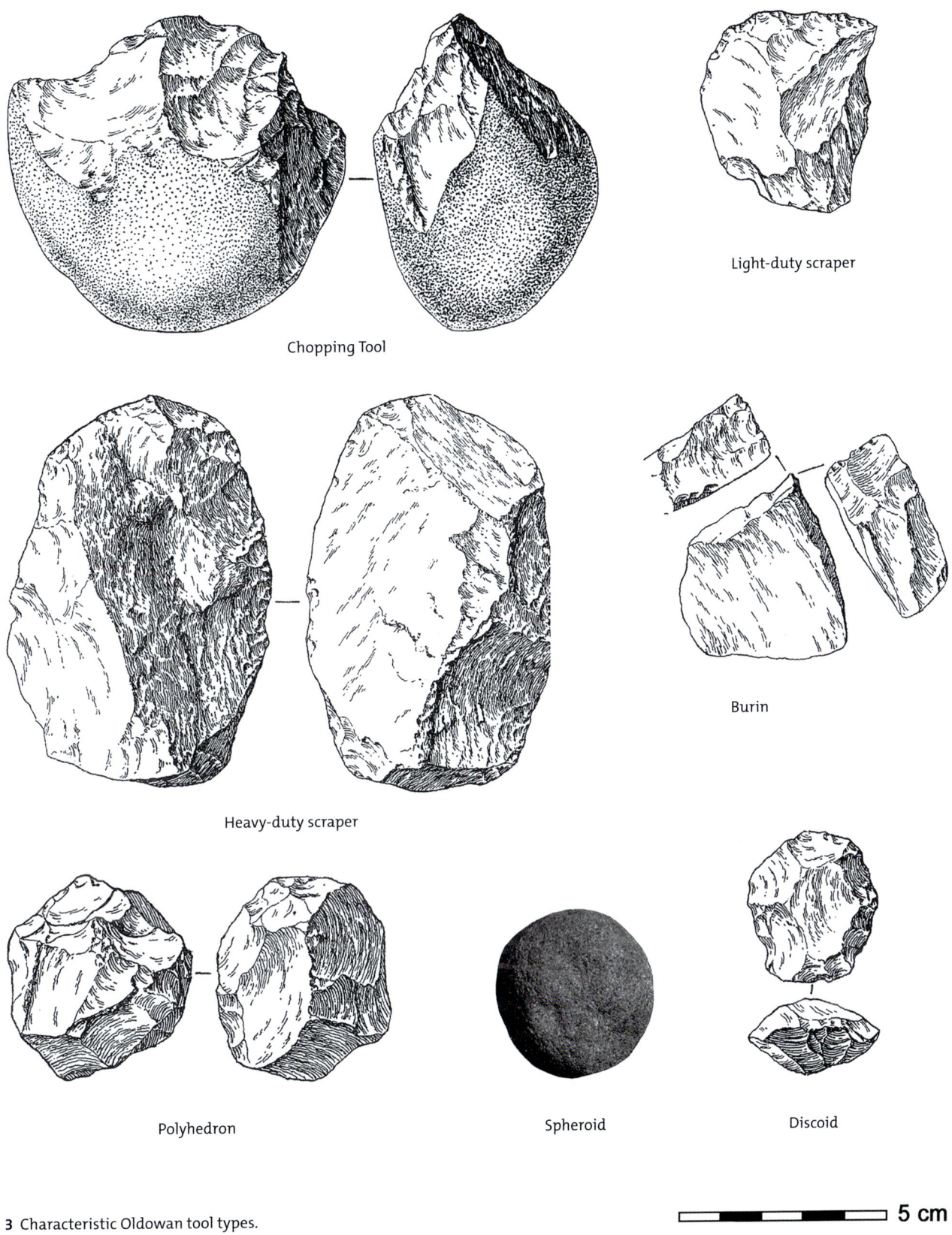

Chopping Tool

Light-duty scraper

Heavy-duty scraper

Burin

Polyhedron

Spheroid

Discoid

3 Characteristic Oldowan tool types.

5 cm

Before the Oldowan: The Lomekwian

Fig. 4

Beginning in 2011, several stone artifacts were found at the Lomekwi 3 site in the West Turkana region of Kenya that are around 3.3 million years old, i.e., around 700,000 years older than the earliest Oldowan artifacts. The manufacturing techniques were simple: as one possibility, the rock to be worked on was held with both hands and struck on an anvil with great force. As another possibility, the piece was held on an anvil with one hand and the other hand struck the piece vertically with a hammerstone (bipolar technique). Evidence of both techniques is very rare in the Oldowan. The new term 'Lomekwian' was proposed since the Lomekwi 3 inventories are different from that of the Oldowan. Typical of the Lomekwian are essentially coarse, often large cores, large flakes, hammerstones, and heavy stone blocks used as anvils.

Characteristics of the Oldowan

Fig. 2

In contrast to the Lomekwian, we find a more targeted basic blank production in the Oldowan, which can even include the serial manufacture of flakes. Usually, the raw piece is held freely in one hand, while the other hand actively knaps off flakes from this core by striking it with the hammerstone. Retouched forms are generally very rare in the Oldowan. Raw materials were usually procured locally, yet research has shown certain rocks with better impact properties were deliberately chosen more frequently than others.

Fig. 3

There are several systems for classifying the Oldowan artifacts. The following description essentially follows the system defined by Mary Leakey in 1971 for the findings from Olduvai Gorge.

- The typical "pebble tools" of the Oldowan are **choppers** and **chopping tools**. For choppers, one single edge is created by knapping off one or more flakes from one side, for chopping tools the edges are prepared by knapping flakes from both sides.
- A **polyhedron** is an angular pebble tool with three or more worked edges that usually intersect. According to Schick and Toth, these are heavily reduced cores.
- A **discoid** is a flat core with a mostly lens-shaped or D-shaped cross-section and a serrated working edge, knapped from both surfaces, that is worked all the way or almost round.
- The **spheroids** include knapped, spherical pebble tools, in which the protruding edges have not been removed or only partially removed. Stone spheres with roughly smoothed surfaces are rare.

- In the context of the Oldowan, the **burin**, a typically Upper Paleolithic and occasionally Middle Paleolithic tool, is a device in which negatives were produced from partially smooth surfaces employing one or more strokes, almost at right angles to the main plane.
- **Heavy-duty scrapers** were often prepared from flat pieces of raw material that are steeply retouched on one or more edges. **Light-duty scrapers**, on the other hand, are made from flakes.

Other characteristic tools include **pointed proto-handaxes**, partially retouched, as a transitional form between choppers and handaxes. Also relevant are **trihedrons**, three-sided retouched pebble tools, as well as **pics**, massive pebble tools that taper towards the top to form relatively narrow points.

The Oldowan in Africa

There is a great deal of agreement about the beginning of the Oldowan around 2.6 million years ago, which generally corresponds to the beginning of the geological age of the Pleistocene. Its end, however, is less clearly defined. For Africa, the term Oldowan is no longer used for inventories that are less than one million years old. Yet Oldowan-typical artifacts were used here and there up until historical times, which is evidence for the effectiveness of these pieces.

There have been numerous attempts to work out a progressive technological development within the Oldowan, but, in the words of Miriam Haidle, "From a technological point of view (...) it is not expedient to classify the early African flake industries into linear groups from simple to complex developing groups of the pre-Oldowan, Oldowan, and developed Oldowan A and B. The Oldowan is better utilized as an overarching, techno-chronological term for a great flake industry tradition between 2.6 and 1.6 million years ago" (original quotation in German). This article, which follows Haidle's approach, includes African Oldowan sites up to an age of 1.5 million years because from this time onward, Acheulean sites are far more common than those of the Oldowan. Within this timeframe, a purely chronological distinction is made between the still relatively rare sites dated to between 2.6 and 2.0 million years ago and the more frequent sites dated to between 1.99 and 1.5 million years before present.

The earliest-known Oldowan finds were discovered in East Africa. The oldest were the finds from Gona in Ethiopia with an age of up to 2.6 million years. Recently, however, artifacts from Ledi-Geraru, also in Ethiopia, were published that are up to 2.61 million years old and thus probably somewhat older than the finds from Gona. Both inventories were primarily made up of simple flakes and cores or 'pebble tools.' We know of other very old Oldowan inventories with an

Fig. 4

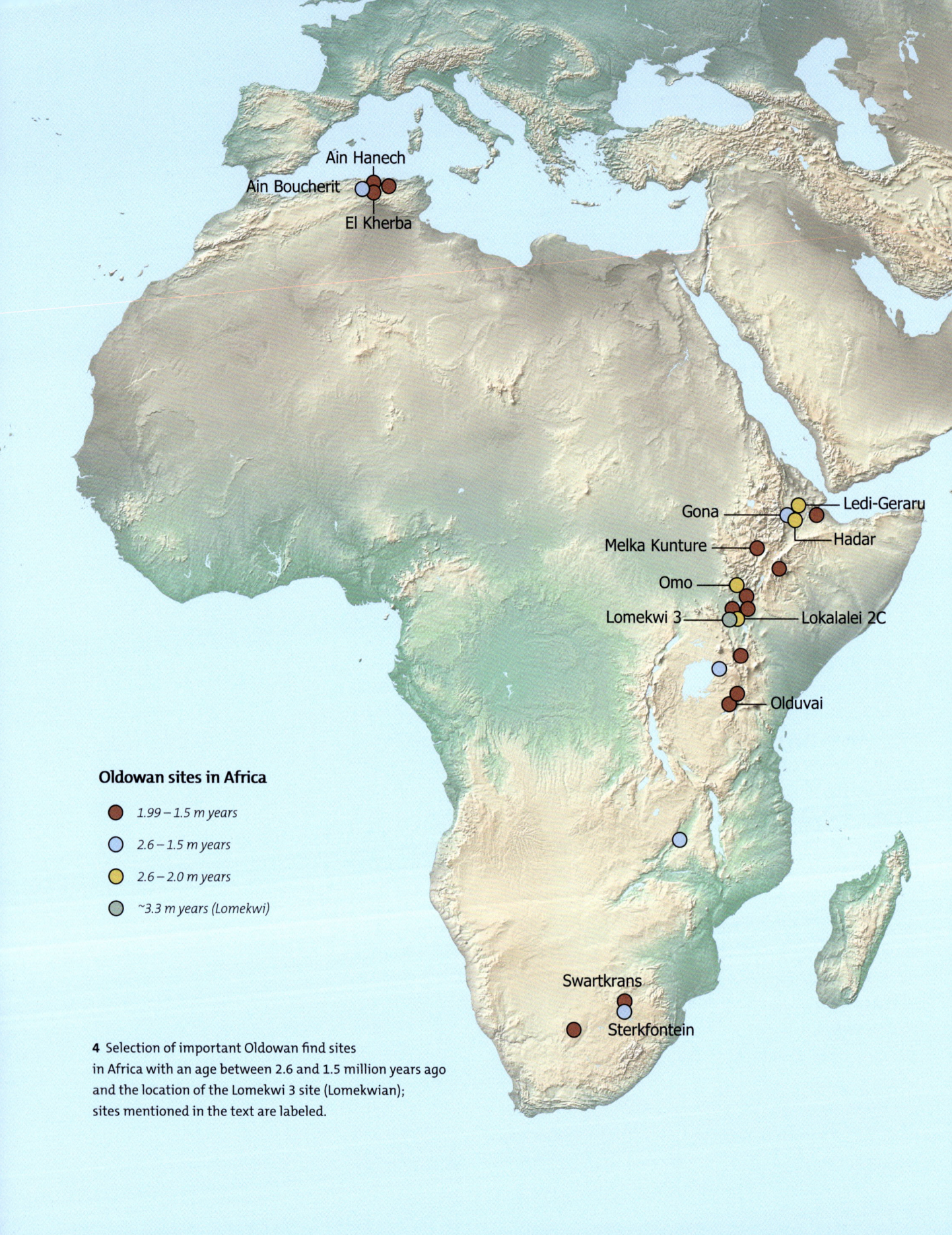

Oldowan sites in Africa

- 🔴 1.99 – 1.5 m years
- 🔵 2.6 – 1.5 m years
- 🟡 2.6 – 2.0 m years
- ⚪ ~3.3 m years (Lomekwi)

Aïn Hanech
Aïn Boucherit
El Kherba

Gona — Ledi-Geraru
Melka Kunture — Hadar
Omo
Lomekwi 3 — Lokalalei 2C
Olduvai

Swartkrans
Sterkfontein

4 Selection of important Oldowan find sites
in Africa with an age between 2.6 and 1.5 million years ago
and the location of the Lomekwi 3 site (Lomekwian);
sites mentioned in the text are labeled.

0 — 1000 km

Scale corresponds to length at the equator

age of up to 2.3 million years in East Africa from the Hadar region and the Shungura formation in the Omo River Valley. The locality Lokalalei 2C in West Turkana in Kenya, which is about 2.34 million years old, deserves special attention. The artifacts testify to further development in the creative drive of their manufacturers. By systematically turning the workpiece several times, it was possible to knap off more than 50 flakes from a single raw material core in a purposeful and directed process – a clear sign of the manufacturer's already high level of planning depth and technical skill.

Fig. 5

The numerous sites in the eponymous Olduvai Gorge are significantly younger, 'only' around 1.8–1.6 million years old. The various sites in the Melka Kunture region in Ethiopia are no more than 1.7 million years old. Evidence of the Oldowan technology was also discovered at very early sites in South Africa, albeit a little later than in East Africa. These include, for example, Sterkfontein with an age of around 2.2 million years, and Swartkrans, 1.7–2.0 million years old. The Oldowan is also very old in North Africa. Two sites in Ain Boucherit in Algeria date back to 2.4 million and 1.9 million years ago, respectively, while the sites Ain Hanech and El-Kherba, also in Algeria, are around 1.8 million years old.

Fig. 4

Expansions: the Oldowan outside of Africa

Although the term Oldowan was and is mainly used for finds from Africa, comparable inventories have also been found outside of Africa, especially in the Levant and in Europe. This article considers sites from the period between 1.8 and 0.78 million years ago, where the end of this period marks the beginning of the Middle Pleistocene. The occurrence of these ancient sites outside of Africa is of great importance for the question of the earliest intercontinental human expansions known under the term 'Out of Africa'.

Fig. 6

The oldest substantiated site of early humans outside of Africa is currently Dmanisi in Georgia with an age of around 1.8 million years. The stone industry is dominated by choppers and chopping tools, cores, and flakes, while retouched pieces are rare. It follows that intercontinental expansions were possible with a simple set of tools and that no Acheulean handaxes, or other bifacial tools, were necessary. On the other hand, since the oldest Oldowan artifacts from Africa are almost a million years older than those from Dmanisi, it can be assumed that the mere possession of a stone tool technology was not enough to facilitate early human expansion out of Africa.

Fig. 7

In Europe, sites in southern and western Europe in particular produced finds that are typical of the Oldowan. The oldest is Pirro Nord in Italy with an age of 1.6 to 1.3 million years. Two sites in the Orce region of southern Spain, namely Barranco León and Fuente Nueva 3, the site Sima del Elefante in the Sierra de

Fig. 8

5 Lokalalei 2C (Kenya). Three circa 2.3 million years old complexes of refits, which document a well-thought-out and organized flake production.

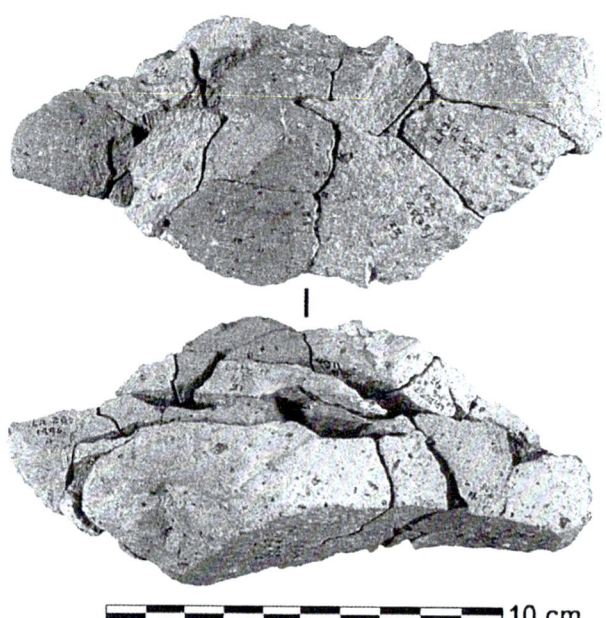

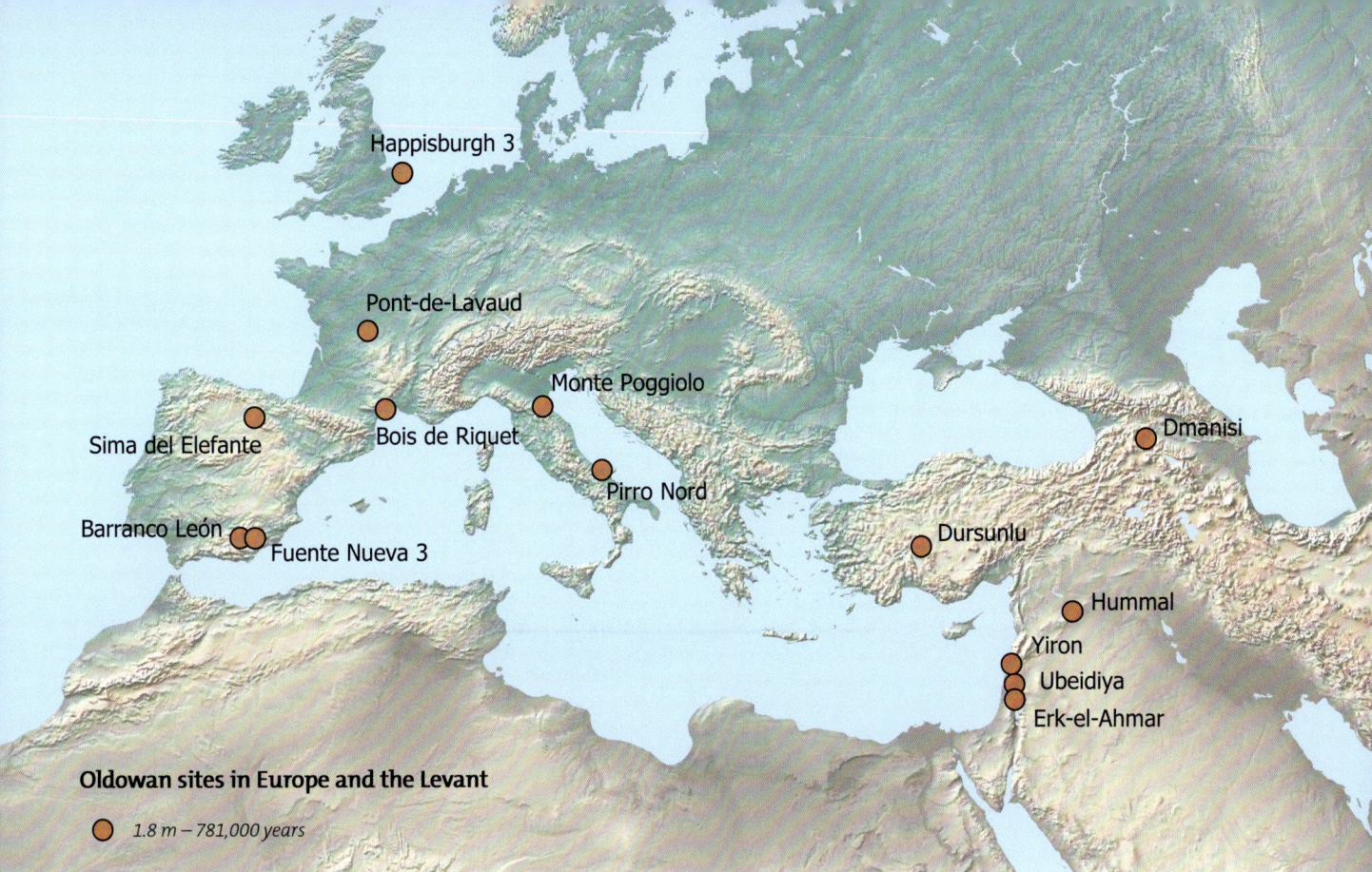

Oldowan sites in Europe and the Levant

● *1.8 m – 781,000 years*

6 Oldowan sites in Europe and the Levant mentioned in the text.

Atapuerca in northern Spain, the French sites Bois de Riquet (Lézignan-la-Cèbe), and Pont-de-Lavaud and Monte Poggiolo in Italy are between 1.4 and around 1.0 million years old. Somewhat surprising, due to the geographical location, is the occurrence of such finds in the British site Happisburgh 3, which are between 970,000 and 780,000 years old.

The Dursunlu site in Anatolia is geographically outside of Europe and no more than 1.1 million years old. Also relevant for the Levant are the Israeli sites Erk-el-Ahmar and Yiron with an age of 1.7 million years and 'Ubeidiya with about 1.4 million years. The age of the Oldowan in the lowest strata of Hummal in Syria is about one million years. Some sites in China and Southeast Asia with stone artifacts typical of Oldowan are only slightly younger than Dmanisi.

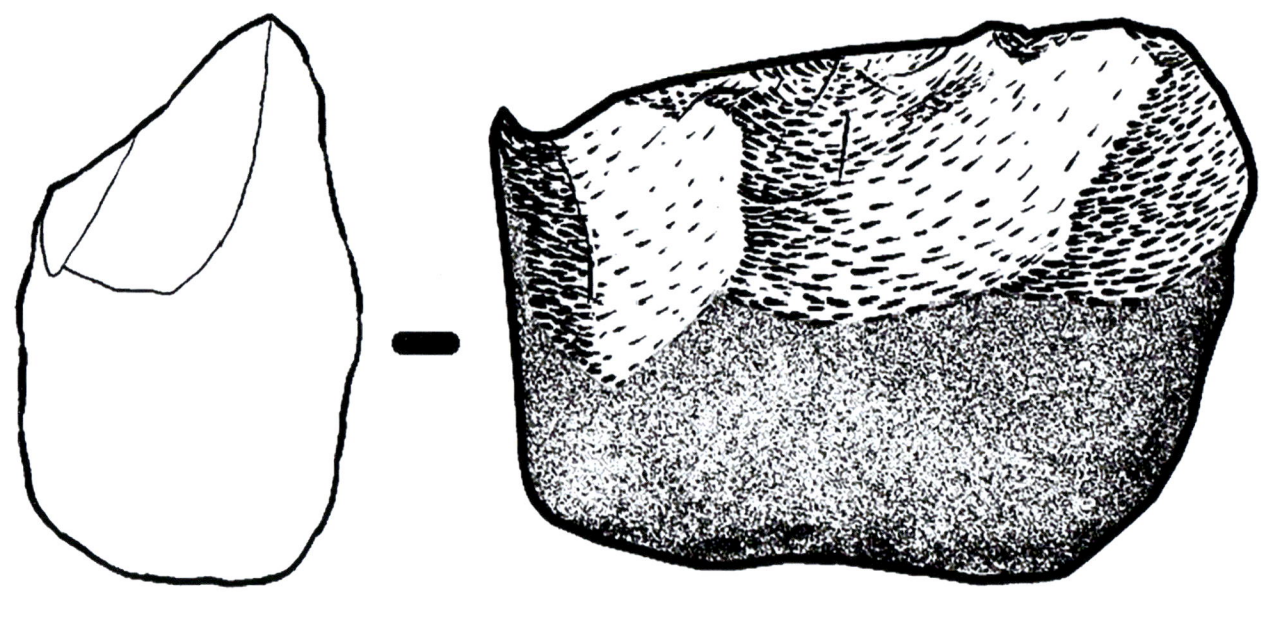

7 Chopper from Dmanisi.

Which humans were responsible for the Oldowan?

Commonly, representatives of the genus *Homo* were considered to be the producers of the oldest stone artifacts, however, late australopithecines also existed during the period assigned to the artifacts from Gona and Ledi-Geraru, and since no fossils were found directly associated with these stone artifacts, the question who produced them remains open. Some evidence that they were produced by a representative of the genus *Homo* was the discovery of the human lower jawbone in Ledi-Geraru, located only about 5 km from the site where the Oldowan artifacts were found. It presumably belongs to *Homo* and, with an age 2.8 million years old, would be the oldest *Homo*-fossil ever found. The finds from Lomekwi 3, on the other hand, date to a period from which no representative of the genus *Homo* is known. The producers must have been representatives of another genus. The numerous human remains from Dmanisi in Georgia are attributed to *Homo erectus* which suggests that the excavated Oldowan artifacts were also produced by these hominins. In Europe and the Levant, *Homo erectus* is also the only species that can be held responsible for the corresponding artifacts.

Oldowan technology and human cognition

The artifacts from Lomekwi 3 already attest to a certain knowledge of the fracture mechanical properties of the stones used by their manufacturers. It can be assumed all the more for the Oldowan technology. This understanding is arguably something that distinguishes even early humans from all animals, including non-human primates such as chimpanzees and bonobos. Another typical human aspect of the production of artifacts is that planning was carried out with foresight. This means that the early humans did not produce the artifacts exclusively for immediate use but were also able to produce tools for some future need for a yet unknown purpose at an unknown place and time. The reduction series from Lokalalei 2C shows that our ancestors finally reached a stage at which humans as artifact producers have left all animals far behind.

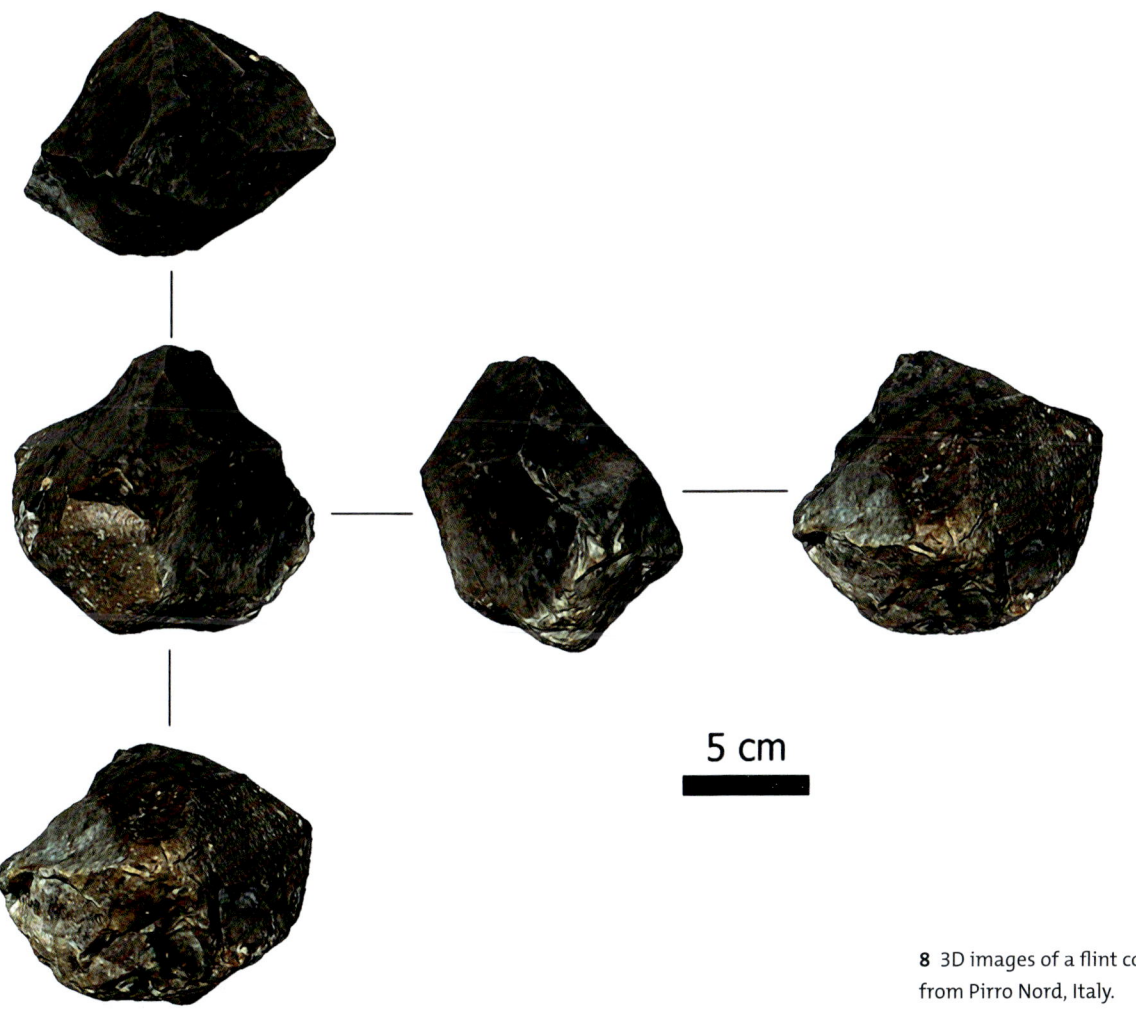

5 cm

8 3D images of a flint core from Pirro Nord, Italy.

Further reading

Braun, D. R./Aldeias, V./Archer, W./Arrowsmith, J. R./Baraki, N./Campisano, C. J./Deino, A. L./DiMaggio, E. N./Dupont-Nivet, G./Engda, B./Feary, D. A./Garello, D. I./Kerfelew, Z./McPherron, S. P./Patterson, D. B./Reeves, J. S./Thompson, J. C./Reed, K. E. 2019 Earliest known Oldowan artifacts at >2.58 Ma from Ledi-Geraru, Ethiopia, highlight early technological diversity. Proceedings of the National Academy of Sciences of the United States of America 116, 2019, 11712–11717.

Delagnes, A./Roche, H. 2005 Late Pliocene hominid knapping skills: The case of Lokalalei 2C, West Turkana, Kenya. Journal of Human Evolution 48, 2005, 435–472.

Haidle, M. N. 2012 Oldowan und andere frühe Geröllgeräte- bzw. Abschlagindustrien. In: H. Floss (ed.), Steinartefakte vom Altpaläolithikum bis in die Neuzeit (Tübingen 2012) 159–166.

Harmand, S. 2007 Economic behaviors and cognitive capacities of early hominins between 2.34 Ma and 0.70 Ma in West Turkana, Kenya. Mitteilungen der Gesellschaft für Urgeschichte 16, 2007, 11–23.

Harmand, S./Lewis, J. E./Feibel, C. S./Lepre, C. J./Prat, S./Lenoble, A./ Boës, X./Quinn, R. L./Brenet, M./Arroyo, A./Taylor, N./Clément, S./Daver, G./Brugal, J.-P./Leakey, L./ Mortlock, R. A./Wright, J. D./Lokorodi, S./Kirwa, C./Kent, D. V./Roche, H. 2015 3.3-million-year-old stone tools from Lomekwi 3, West Turkana, Kenya. Nature 521, 2015, 310–315.

Leakey, M. D. 1971 Olduvai Gorge. Excavations in Beds I & II 1960-1963 (Cambridge 1971).

Schick, K./Toth, N. 2006 An overview of the Oldowan industrial complex: the sites and the nature of their evidence. In: N. Toth/K. Schick (eds.), The Oldowan: case studies into the earliest Stone Age (Gosport 2006) 3–42.

Semaw, S./Rogers, M. J./Quade, J./Renne, P. R./Butler, R. F./Dominguez-Rodrigo, M./ Stout, D./Hart, W. S./Pickering, T./Simpson, S. W. 2003 2.6-Million-year-old stone tools and associated bones from OGS-6 and OGS-7, Gona, Afar, Ethiopia. Journal of Human Evolution 45, 2003, 169–177.

Australopithecus africanus

○ *Australopithecus africanus*

0   1000 km

Discovery

The first fossil was discovered in 1924 by Raymond Dart in a fossil collection in Taung, South Africa. It was an almost complete skull of a child with a few teeth and led to the first initial description of *Australopithecus*.

Sites

South Africa: Sterkfontein, Makapangsgat, Taung.

Finds

Skull and lower jaw bone of a child, fossilized cranial imprint, skull ("Mrs. Ples"), isolated teeth and lower jaw bone fragments

Age

3.0–2.1 million years.

Brain size

450–550 cm^3.

Characteristics

Australopithecus africanus has only been found in southern Africa. They have a slightly sloping, protruding face, a fleeing forehead, but pronounced brow ridges above the eyes. The position of the occipital hole is more similar to humans than African great apes, which is why it can be assumed that they were constantly moving around on two legs. Representatives of the species *Australopithecus africanus* were estimated to be 1.30 m tall and weighed around 30–40 kg. Since they were omnivores, their diet probably included not only leaves, tubers, roots, lichens, tree bark, and seeds, but also meat. Their habitat were wooded areas near rivers, so-called gallery forests.

Facial reconstruction

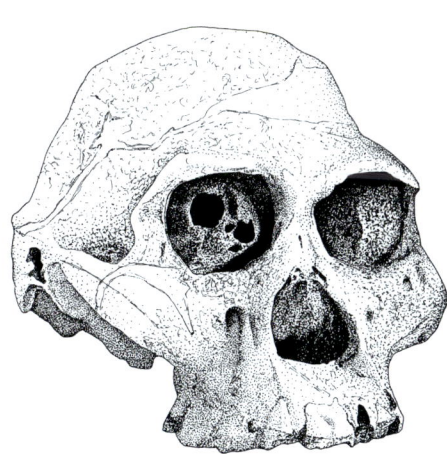

"Mrs. Ples" from Sterkfontein, South Africa (STS 5)

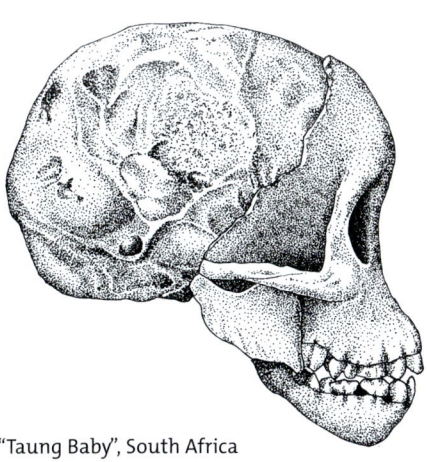

Skull of the "Taung Baby", South Africa

Liane Giemsch

From Africa around the world: the Acheulean

The oldest known tools of early humans, processed pebbles and sharp-edged flakes, were documented in Africa, the cradle of humankind, at the 3.3-million-year-old Lomekwi 3 site in Kenya and various other sites from around 2.6 million years ago (see article by Bolus in this volume). Around 1.76 million years ago, stone tool technology changed significantly and artifacts worked on two surfaces, so-called bifaces, appeared. In addition to the pebble cores and chopping tools of the Oldowan, the recognizable handax now appears as the type specimen of the so-called Acheulean culture. It is a much more complex device than the previous stone tools and was mostly fashioned from basalt. These assemblages are named after the site Saint Acheul in northern France, where Jacques Boucher de Perthes (1788–1868) collected handaxes as early as the 1830s and interpreted them as human products. The researcher couple Mary and Louis Leakey carried out the first modern archeological excavations of the Acheulean in Africa at the Olorgesailie site in the years 1943–1947. Further work in Africa followed at Olduvai Gorge, Kalambo Falls, and Peninj, among others. Research into the early Pleistocene cultures of East Africa received considerable impetus from the discovery of fossil hominins in East Africa and the intensification of research into primate behavior, among others by Jane Goodall.

The Acheulean is the younger phase of the African Early Stone Age. It marks an important stage in the history of human technology and behavioral evolution and, with around 1.5 million years, has a similar duration as the previous Oldowan. Bifacial reduction or retouch on both surfaces of a core is also referred to as Mode II, based on the Grahame Clarke classification.

1 Acheulean basalt handax from the Makuyuni site in Tanzania.

2 Production scheme of an Acheulean handax.

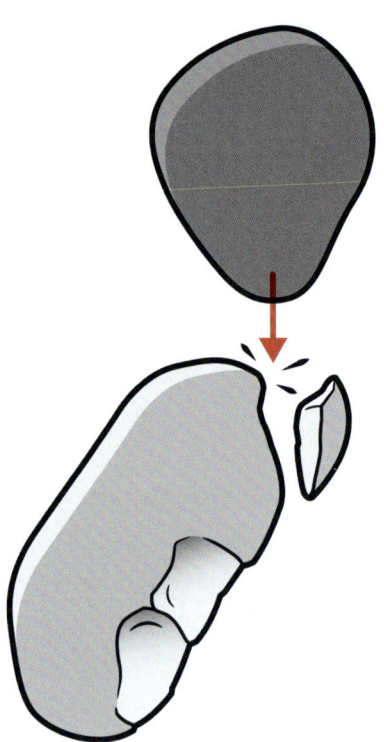

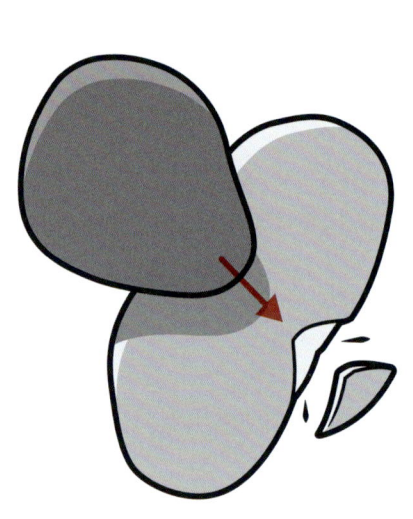

1.

2.

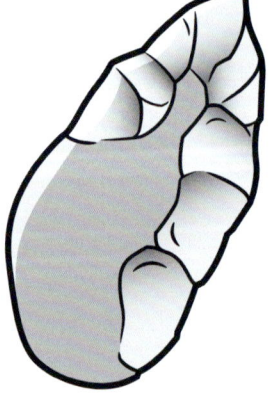

3.

4.

According to the current state of research, the Acheulean appears for the first time in East Africa. The oldest sites are Kokiselei 4 in West Turkana, Kenya, Konso-Gardula and Gona in Ethiopia, circa 1.76 million years old, as well as Peninj on Lake Natron, Tanzania, which is between 1.5 to 1.1 million years old. Until the end of the Middle Pleistocene around 130,000 years ago, the industry was present throughout western Asia south of the mountains, including India. The first Paleolithic handaxes discovered outside of Africa were found in 'Ubeidiya, Israel, and date to 1.4 million years before present.

Characteristics of the Acheulean

The Acheulean is divided into three phases: an early (about 1.76 to 1 million years before present), a middle (about 1 to 0.6 million years before present), and a late phase (about 0.6 to 0, 3 million years before present). In the last phase of the Oldowan culture, the first proto-handaxes appeared, heralding the transition to the Acheulean. This transition appears to be rapid and very few transitional assemblages exist. The cognitive processes associated with the conception of the typical Acheulean tool forms and the techniques for producing handaxes are very different from those used in the Oldowan. While Oldowan flakes were generally knapped from fist-sized pebbles, the Acheulean toolmakers preferred to knap off very large flakes from large boulders or cores in order to continue to process them into handaxes, picks, or cleavers. Experiments have shown that in *Fig. 2* contrast to the striking technique practiced in the Oldowan, in which the workpiece was held freely in the hand, the large flakes of the Acheulean were made by placing the piece onto a stone or wooden anvil or simply by placing the core on the ground. This knapping-off strategy required a lot more strength as well as excellent coordination and precision.

In the early Acheulean, handaxes included a round, thick end and a narrow, pointed end. The middle phase of the Acheulean is not a clearly defined cultural/technological stage, either chronologically or technologically. The pieces show moderately refined biface production and, from about one million years ago, oval, triangular, and other handax forms with a more predetermined shape and an increasing emphasis on symmetry and balance. The late Acheulean is ascribed in Africa from between 600,000 to 500,000 years ago up into the *Fig. 3* Middle Stone Age around 300,000 years ago. The bifaces undoubtedly became finer (thinner, more symmetrical, and with more flake negatives), which may have been possible due to the introduction of the soft knapping technique. In contrast to the hard striking technique with a hammerstone, the soft knapping technique requires an organic mallet made of antler or bone to work the stone. This makes it possible to knap off significantly finer flakes and to produce

3 The sophistication of handax production over time can be seen at the site in Konso (Ethiopia).

Pictured from left to right are sets of two handaxes each: approx. 1.75 million years, 1.6 million years, 1.25 million years, and 0.85 million years old (above the dorsal, below the ventral surfaces). The degree of preparation varies from nearly unifacial (left) to extensive bifacial processing (right).

correspondingly finer tools. Furthermore, numerous sites in Africa already exhibit evidence for the use of the so-called Levallois technique for the extraction of much finer and thinner flakes, which heralded the transition to the Middle Stone Age industries around 300,000 years ago. The abandonment of handaxes and cleavers in favor of smaller flaked tools in the Middle Stone Age represented the replacement of tools that were freely held in the hand with hafted pieces. It signals a profound technological reorganization during the transition from the Acheulean to the Middle Stone Age, which is associated with the appearance of *Homo sapiens.* The earliest Middle Stone Age artifacts from the Baringo site date back to 284,000 years ago. The Acheulean disappears in most regions of Africa around 200,000 years ago.

Fig. 4   The Acheulean is characterized by two special tool shapes: the handax and the cleaver. Handaxes, for some THE symbol of the Paleolithic, are large (> 10 cm) tools made from a pebble or large flake and reduced into a teardrop or triangular shape, with one narrow pointed end and another wider and often rounded end. Cleavers are similar in size, but instead of a pointed end, they have a wide

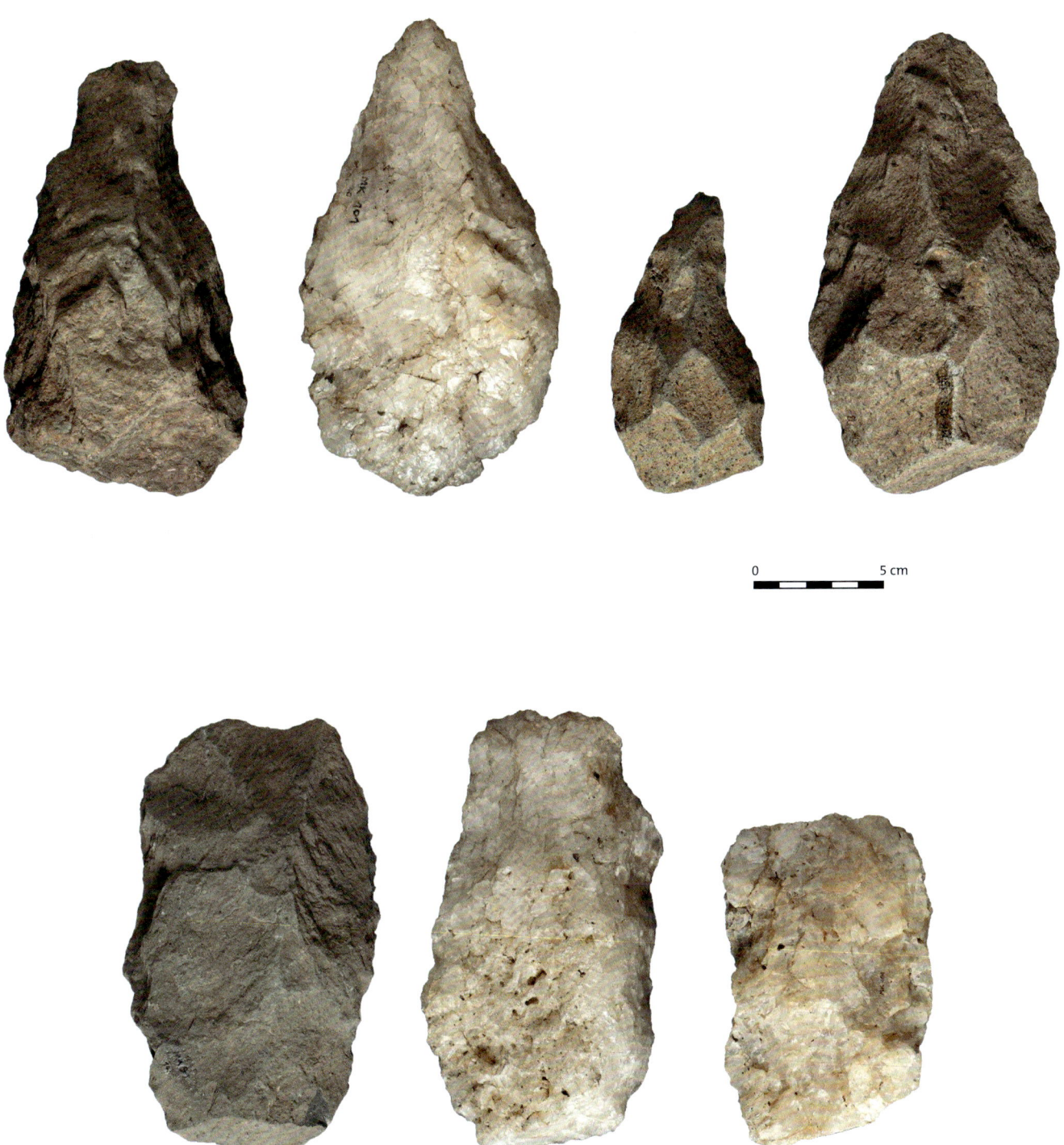

4 Handaxes and cleavers (below) from the Makuyuni site at Lake Manyara in Tanzania.
The basalt and quartz finds are about 600,000–400,000 years old.

5 Facial reconstruction of a *Homo erectus*.

cutting edge across the longitudinal axis of the tool. Both handaxes and cleavers are usually bifacial, that is, worked on both surfaces. Picks are similar to hand-axes, but thicker overall and often triangular in cross-section. There is evidence that these large tools were used effectively to dissect large animals such as elephants and rhinos but were also used to dig in the earth for woodworking. Due to the similarities in the mode of production and morphological transition phenomena, cleavers are also understood as broad-edged handaxes. They were probably used in almost the same functional context. Since the handax also occurs beyond the Acheulean in the cultures of the Middle Paleolithic, it accompanies human history over an immense period of circa 1.7 million years!

Way of life

Overall, little is known about the way of life of the people of the Acheulean. While even spears made of wood were preserved in the advanced Middle Pleistocene, very little is known for its early phase. At a few sites such as Olorgesailie in Kenya, artifacts could be documented in the context of butchered elephant remains, which are evidence for the hunt for large mammals. Many of the animal bones were broken open to allow for bone marrow extraction. Aside from the use of stone tools, plant remains from Kalambo Falls in Tanzania show that humans probably used plant resources as tools and food, as was the case in Gesher Benot Ya'aqov in Israel, where stones were found with dimples that were created through the repeated cracking of nuts. It should also be noted that the maximum transport distance of the raw materials increased from 15 km in the Oldowan and early Acheulean periods to 45 km in the Middle Acheulean. Evidence of early fire use from 1.5 million years ago is available from various sites such as Swartkrans (South Africa), Koobi Fora and Chesowanja (both Kenya) (see the article "Fire" by Giemsch in this volume).

Which hominins are responsible for the Acheulean?

The Acheulean probably emerged with the first representatives of *Homo ergaster* or archaic *Homo erectus*. From the Middle Pleistocene onwards, *Homo heidelbergensis* probably also used this technology. Examples are known from Tighénif *Fig. 5* (Algeria), Olduvai (Tanzania), Melka Kunture (Ethiopia), Ndutu (Tanzania), and Olorgesailie (Kenya). Early representatives of *Homo sapiens*, for example from Jebel Irhoud (Morocco), Ngaloba (Tanzania), and Haua Fteah (Libya) can be linked to the Middle Stone Age technology that followed the Acheulean. This leads to the assumption that the Acheulean-to-Middle Stone Age transition about 250,000–300,000 years ago corresponds to the species change from *Homo heidelbergensis* to *Homo sapiens*.

What role do environmental changes play in the emergence and evolution of *Homo ergaster* and the Acheulean? Most of the Oldowan sites in Africa are located on the banks of a lake or in floodplains, and mainly in lower areas of the rift, while the toolmakers of the Acheulean occupied a wider variety of habitats, including drier and higher-lying areas. They were also likely the first hominins to venture out of Africa in large numbers, although Acheulean technology wasn't widespread in Eurasia until much later, after one million years. Significant changes in the global climate took place in East Africa by 1.9–1.7 million years. There was increased drought and grassland expansion. While it is likely that these environmental changes and increased seasonality and variability played a significant role in the emergence of *Homo erectus*, the Acheulean, and possibly in the changing adaptations of the Middle and Late Acheulean, it is still not clear which specific selective factors triggered these biological and technological changes.

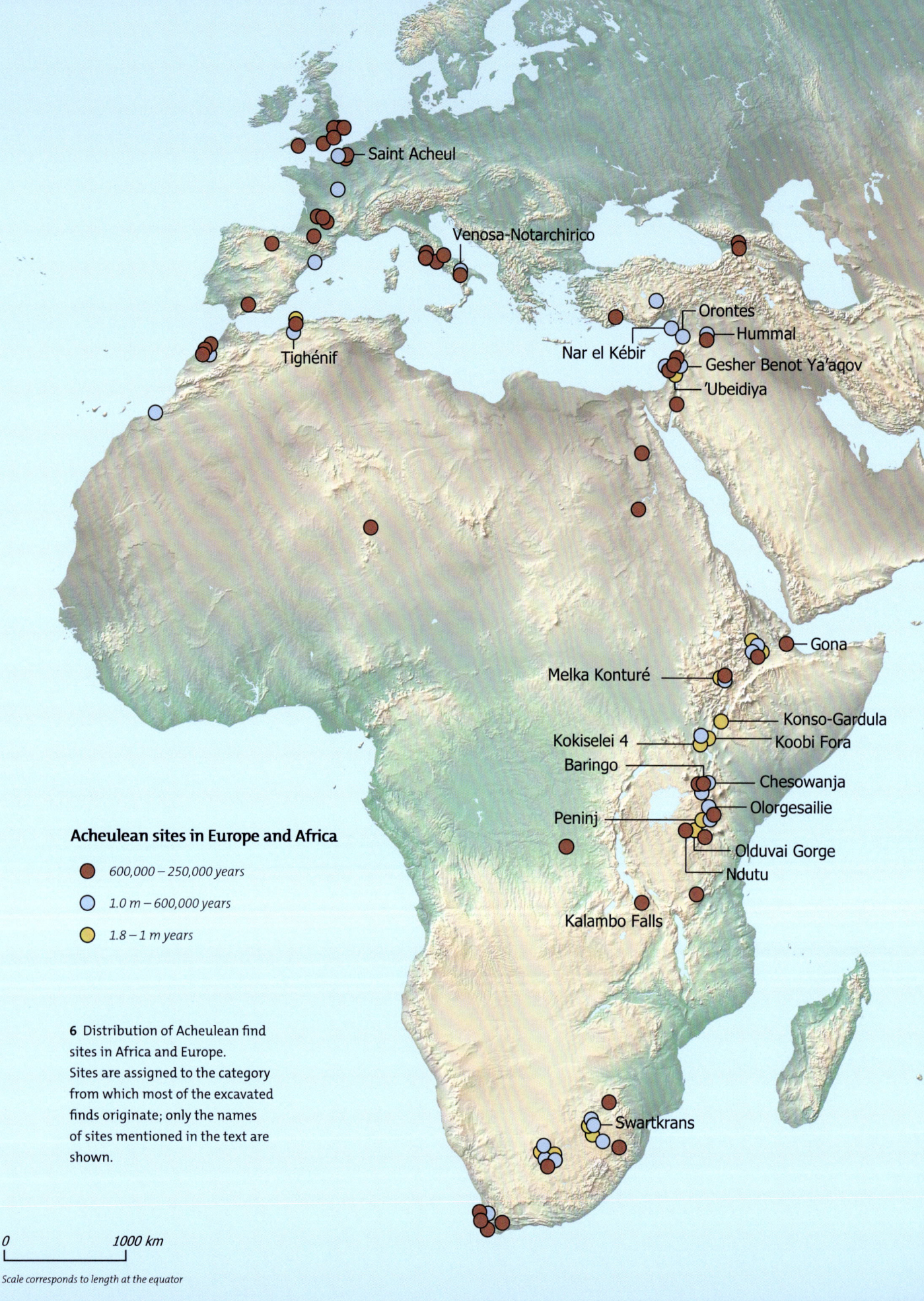

Acheulean sites in Europe and Africa

- 🔴 600,000 – 250,000 years
- ⚪ 1.0 m – 600,000 years
- 🟡 1.8 – 1 m years

6 Distribution of Acheulean find sites in Africa and Europe. Sites are assigned to the category from which most of the excavated finds originate; only the names of sites mentioned in the text are shown.

Saint Acheul

Venosa-Notarchirico

Orontes

Hummal

Nar el Kébir

Gesher Benot Ya'aqov

'Ubeidiya

Tighénif

Gona

Melka Konturé

Konso-Gardula

Koobi Fora

Kokiselei 4

Baringo

Chesowanja

Olorgesailie

Peninj

Olduvai Gorge

Ndutu

Kalambo Falls

Swartkrans

0   1000 km

Scale corresponds to length at the equator

Spread of the Acheulean culture

In addition to Africa, the Acheulean technology is also documented in large parts of Europe and Asia. The oldest unambiguous handaxes in the Middle East were discovered in 'Ubeidiya with an age between 1.4 and 1 million years before present. The pieces from Hummal, Sitt Markho (Nar el Kébir), and Khattab (Orontes) in Syria are of comparable age. In Europe, there are only a few sites in Spain, Italy, and southern and central France that have delivered proto-handaxes or poorly preserved handaxes that are more than 780,000 years old. Another expansion wave could be documented through the 800,000-year-old Gesher Benot Ya'aqov site in Israel. In addition to basalt handaxes, cleavers that first appeared in Africa around a million years ago, for example, at the Olorgesailie site in Kenya, were also found at the site in Israel. The first handax inventories in Europe date to between 900,000 and 500,000 years ago. In Venosa-Notarchirico, southern Italy, the industry occurs together with the remains of forest elephants. The geographical bottleneck of the Middle East on one side and the Strait of Gibraltar on the other side are both conceivable as diffusion routes to Europe and Asia.

Fig. 6

Conclusion

The Acheulean is perhaps the longest-lived technological tradition in human history. In Africa, it extends from 1.7 to 0.3 million years and corresponds roughly to the time in which *Homo erectus (Homo ergaster)* and *Homo heidelbergensis* lived there. In contrast to the earlier Oldowan technology, Acheulean tools—mostly handaxes, cleavers, and picks—were formed from large boulders and flakes and became increasingly standardized. The long duration of the Acheulean for over 1.4 million years is proof of the success of this technology in different habitats, altitudes, and environments, but also its conservative character since its tradition was passed on among highly mobile hominin groups with small populations over thousands of generations. Although there are differences between the early and late Acheulean industry, several researchers see technological inertia in the stone tool technology of the Acheulean, which also characterizes the previous Oldowan. Nevertheless, the makers of these tools experienced major changes through the use of other technologies (for example the use of wood, bone, and fire), strategic land use, and way of life (for example group size, organization, type of cultural transmission). Although technologically almost static, the symmetry and standardization of the Acheulean formed the basis for the later development of symbolism and language.

Further reading

Beyene, Y. /Katoh, Sh./WoldeGabriel, G./Hart, W. K.; Uto, K./Sudo, M./Kondo, M./Hyodo, M./Renne, P. R./ Suwa, G./Asfaw, B. 2013 The characteristics and chronology of the earliest Acheulean at Konso, Ethiopia. Proceedings of the National Academy of Sciences of the United States of America 110, 2013, 1584–1591.

Clarke, G. 1969 World prehistory (Cambridge 1969).

Isaac, G. L. 1977 Olorgesailie. Archeological studies of a Middle Pleistocene lake basin in Kenya (Chicago, London 1977).

Le Tensorer, J.-M. 2012 Faustkeile. In: H. Floss (ed.), Steinartefakte. Vom Altpaläolithikum bis in die Neuzeit (Tübingen 2012) 209–218.

Leakey, M. D. 1971 Olduvai Gorge Volume 3. Excavations in Beds I and II, 1960–1963. Olduvai Gorge 3 (Cambridge 1971).

Lepre, C. J./Roche, H./Kent, D. V./Harmand, S./Quinn, Rh. L./Brugal, J.-Ph./Texier, P.-J./ Lenoble, A./Feibel, C. S. 2011 An earlier origin for the Acheulian. Nature 477, 2011, 82–85.

McBrearty, S./Brooks, A. S. 2000 The revolution that wasn't: a new interpretation of the origin of modern human behavior. Journal of Human Evolution 39, 2000, 453–563.

Moncel, M.-H./Schreve, D. 2016 The Acheulean in Europe: origins, evolution and dispersal. Quaternary International 411, Part B, 2016, 1–8.

Sahnouni, M. 2013 The African Acheulean. An archaeological summary. In: P. Mitchell/ P. Lane (eds.), The Oxford Handbook of African archaeology (Oxford 2013) 307–323.

Saragusti, I./Goren-Inbar, N. 2001 The biface assemblage from Gesher Benot Ya'aqov, Israel: illuminating patterns in "Out of Africa" dispersal. Quaternary International 75, 2001, 85–89.

Paranthropus boisei

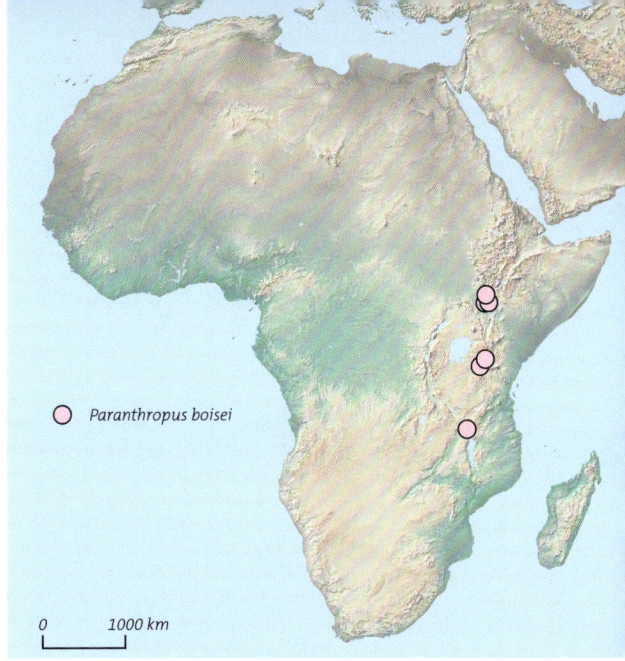

○ *Paranthropus boisei*

0   1000 km

Discovery

The first remains of a *Paranthropus boisei*, a skull including teeth, were discovered by Mary Leakey in 1959 at Olduvai Gorge in Tanzania.

Sites

Tanzania: Peninj, Olduvai.

Malawi: Malema.

Kenya: Koobi Fora, Nachukui, Chesowanja.

Ethiopia: Konso, Shungura.

Finds

Skull, teeth, lower jaw bone, an ankle joint, thumb bones, and lower leg bones.

Age

2.3–1.4 million years.

Brain size

475–545 cm³.

Characteristics

The skulls of *Paranthropus boisei* are large and have a long face with a powerful lower jaw. Special features are the broad cheek-bones and the very large molars. The strong masticatory muscles were attached to a bony sagittal crest. As with *Paranthropus robustus*, the diet was probably limited to seeds, roots, and tubers, supplemented by fruits, leaves, and occasional insects. It is not yet clear whether *Paranthropus boisei* also ate meat.

Facial reconstruction

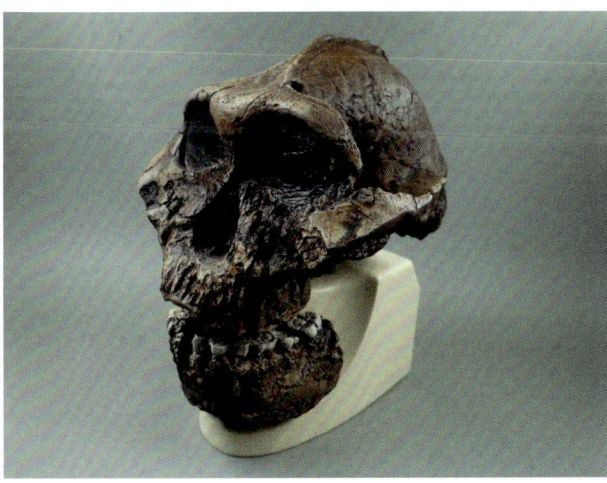

Skull KNM-ER 406 from Koobi Fora, Kenya

Liane Giemsch

Early human use of fire

Handling of fire has become something we take for granted today. Humans use it in a variety of ways, and it has become a constant, always available companion. We rely on and are dependent on it. Even if, in our modern society, fire is often hidden from view, almost all the achievements of the industrial age are based on it: metals, glass, plastics, ceramics, power generation, combustion engines, and rocket propulsion. Without the power of fire, our civilization would not exist in this form. Its use was a qualitative leap for our ancestors, and control over it marks a clear boundary between animals and humans. No other distinguishing criterion achieves this exclusivity in the discussion about our delineation from animals. The manipulation and creation of fire is an exclusively human trait that has become universal within our *Homo* species. The ability to control fire is a crucial trait of human culture and has likely influenced both the physical and cultural development of our lineage. Fire has fundamentally changed our relationship with the world. But when did humans begin to use fire, and what are the many advantages of this cultural innovation?

The benefits of fire for those who were brave enough to harness it are many. The use of fire as a heat source extended the natural range of humans and made it possible for them to colonize more northern latitudes and higher altitudes. For the first time, fire also provided an effective deterrent against dangerous predators and enabled hominins to occupy caves and drive away other competing cave dwellers such as hyenas and bears. The smoke also kept annoying flies and swarming mosquitos away. Fire also provided comfort as it was an effective way of clearing sleeping areas of the parasites that lived in the old grass beds.

Fig. 2

1 The control of fire has become something we take for granted today.

Source of heat

Preservation of food
through smoke

Hunting and cultivating
the landscape

Cooking

Communication

Light source

Smoke as protection
against mosquitos

Protection from predators

Cleaning of sleeping places
using heat

Technical improvement

As a further advantage, fire brought light into the darkness and thus lengthened the day. It provided light for working and kept warm in the cool night. The extra time gained in this way could be used for social interaction, exchange of information, and creative processes. In general, the social component of fire as a central spot for coming together was certainly important. In addition to the social nature of eating together, working around a fire led to an expansion of communication and solidarity within the group, and encouraged cultural and technological advances. Sitting around the fire in the evenings, during and after meals, and the associated exchange of stories and experiences strengthened memories, imagination, and empathy for the thoughts of others. The use of fire surely also led to a closer social structure within the groups since, in addition to the technical challenges, it was also necessary to organize a regular supply of fuel to keep the fires burning.

The use of fire also facilitated technical improvements and innovations. By applying fire (and heat), it was possible to improve the material properties of wood and stone. Through heating (tempering), some rocks became easier to split, making it easier to produce certain tools; wooden lance tips could be hardened by fire. Later, it became possible to manufacture completely new materials such as birch pitch, which was used as an adhesive. Fire was often useful during hunting. With its help, it was possible to create panic among prey animals or corner them so that they could be hunted more easily. It could also be used to cultivate the landscape and thus promote the growth of preferred edible plants for prey species or the hominins themselves. Heat and smoke preserved food through drying and smoking, respectively. This is still of great importance for hunting societies today, to preserve food as a reserve for hard times and thus compensate for bad hunts.

However, the ability to cook food was probably the most important advantage of using fire and is viewed by many scientists as a decisive step in human evolution. Cooking with fire significantly expanded the range of foods that hominins could consume. The heat decomposed poisonous substances in plant foods and eliminated parasites. Fire also resulted in several energetic advantages: the reduced digestive effort when consuming cooked food as opposed to raw food led to a significant reduction in expended energy and time (more on this in the article by Bruch/Hahn in this volume).

2 The many advantages of using fire.

The long road to using fire

Imagine, what were the first steps in harnessing fire? Using fire does not necessarily mean being able to produce it. In general, the process can be summarized in the following steps: 1. Getting used to natural fire, 2. Using fire, 3. Maintaining fire, and 4. Producing fire. In the beginning, it was certainly necessary to overcome the initial reflex to escape or run away when confronted with a natural fire that was ignited through a lightning strike, volcanic activity, or, more rarely, by the spontaneous combustion of coal, oil shale, or other concentrations of organic plant matter. Like today's chimpanzees, early humans developed the ability to face a burning landscape calmly without panic. The next insights included recognizing and making use of the positive consequences of fire, i.e., after a bushfire, such as the easier acquisition of formerly hidden fruits, seeds, or tubers, as well as (lightly cooked) small animals that perished in the fire. In addition, the reduced vegetation cover caused by bushfires made locomotion and the early detection of dangerous predators easier. This passive use of fire probably eventually led to active use. Initially, a naturally burning fire was fed with additional fuel to artificially prolong its presence at the site of origin so that one could warm oneself or cook something. This likely evolved into the ability to transport fire from its place of origin to another location. Eventually, people discovered, probably through a combination of lucky chance and experimentation, that they could make fire themselves whenever and wherever they wanted. This mastery gave early humans profound freedom to control their environment, cook their food, and produce new materials at will. It is generally assumed that these transitions took place in landscapes in which lightning-related fires prevailed or in zones of long-term active volcanism (for example the African Rift Valley). It is possible, that life in fire-prone environments led to an adaptation in hominins that eventually taught them to use fire to their advantage.

What about the evidence?

The earliest evidence suggested for the use of fire by early humans is not archeological but physiological. It is postulated that eating easily digestible cooked food may have been responsible for the shortening of the human intestine and thus the redirection of the calories saved during digestion into the brain, which ultimately led to an increase in brain size as documented in the fossil record with the appearance of *Homo erectus* from around 1.9 million years ago. However, the extra calories needed for larger brains could just as easily come from high-energy bone marrow. At this time, hominins had already mastered the ability to break open skulls and long bones of their prey with large stones to reach the precious bone marrow.

Zhoukoudian ☗

Gesher Benot ☗
Ya'aqov

☗ Gadeb
☗ Koobi Fora
Chesowanja ☗
☗ Olorgesailie

☗ Kalambo Falls

☗ Swartkrans
Wonderwerk ☗ ☗ Florisbad

0   1000 km

Scale corresponds to length at the equator

The earliest archeological evidence of fire use in the form of thermally altered sediments, stone artifacts, or bones was discovered in Africa. The oldest evidence is 1.5 million years old and comes from Koobi Fora in Kenya. Other sites include Chesowanja in Kenya and Gadeb in Ethiopia, as well as the one-million-year-old cave sites Swartkrans and Wonderwerk in South Africa, where an analysis of the sediments has shown that the burned bones were certainly not caused by bush fires, and the site Olorgesailie in Kenya. The oldest evidence for something resembling a stove on which food was cooked is around 790,000 years old and comes from the Gesher Benot Ya'akov site in Israel. Possibly similarly old evidence for the use of fire was discovered in the Zhoukoudian Cave in China. From 400,000 years ago, evidence for use of fire increased significantly. Clear evidence of the production of fire is circa 30,000 years old, yet newly discovered traces on Neanderthal stone tools have also been interpreted as an indication for the early production of fire.

3 Important sites with early fire use in Africa and the Middle East with an age between 1.5 million and 125,000 years.

Further reading

Clark, J. D./Harris, J. W. K. 1985 Fire and its roles in early hominid lifeways. The African Archaeological Review 3, 1985, 3–27.

Goren-Inbar, N./Alperson, N./Kislev, M. E./Simchoni, O./Melamed, Y./Ben-Nun, A./Werker, E. 2004 Evidence of hominin control of fire at Gesher Benot Ya'aqov, Israel. Science 304, 725, 2004, 725–727.

Pruetz, J. D./Herzog, N. M. 2017 Savanna chimpanzees at Fongoli, Senegal, navigate a fire landscape. Current Anthropology, 58, 2017, 16; 337–350.

Sorensen, A. C. 2019 The uncertain origins of fire-making by humans: the state of the art and smouldering questions. Die ungewissen Anfänge der Feuerherstellung durch den Menschen: Forschungsstand und schwelende Fragen. Mitteilungen der Gesellschaft für Urgeschichte 28, 2019, 11–50.

Wrangham, R. W. 2009 Catching fire: how cooking made us human (New York 2009).

Homo rudolfensis

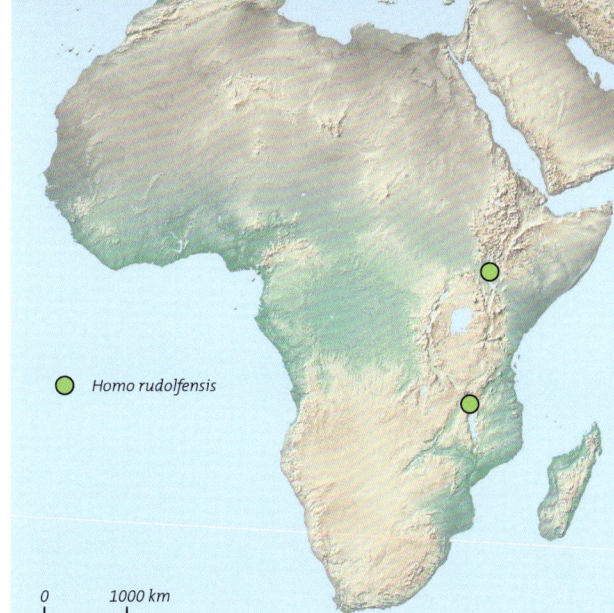

Discovery

The first fossil of *Homo rudolfensis* was discovered by Bernard Ngeneo in 1972 near Lake Turkana. It was a badly fragmented skull.

Sites

Kenya: Koobi-Fora-Formation.

Malawi: Uraha.

Finds

Multiple preserved skulls, lower jaw bone, teeth, thigh bone without articular surfaces, upper portion of a lower arm bone, pelvic bones, shin bone.

Age

2.5–1.8 million years.

Brain size

750–752 cm³.

Characteristics

Homo rudolfensis is the oldest species of the genus *Homo*. The shapes of their leg and pelvic bones indicate that they probably walked bipedally more often and longer than any other species before them. The slightly curved shape of their teeth is also very similar to younger *Homo* species. It is estimated that they were 1.5 m tall and weighed between 45 and 50 kg. The proportion of plants in their diet was large. They mainly ate leaves, seeds, and fruits from trees. While *Homo rudolfensis* is believed to have been able to make and use tools, no tools have yet been found that are directly associated with them.

Facial reconstruction

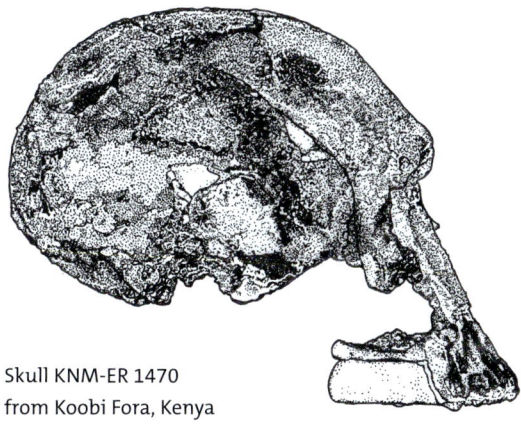

Skull KNM-ER 1470
from Koobi Fora, Kenya

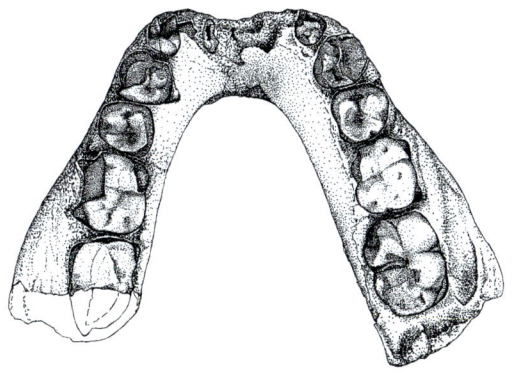

Lower jaw UR 501 from Uraha, Malawi

Angela A. Bruch and Karen Hahn

Raw or roasted?
How fire changed what's on the menu

Wild plants have always played an important role in human nutrition. Although fossil remains of plants are rarely preserved in archeological contexts, it is undisputed that early humans also pursued a flexible and diverse nutritional strategy to meet their needs for energy and nutrients. This diet of meat and edible plant parts improved and became more effective throughout the course of human history through the use of fire and technological advances. At the beginning of human history, however, early humans did not use fire to process their food. Nevertheless, it can be assumed that early humans were able to implement simple methods of obtaining food such as peeling fruits or roots, cracking nuts, or digging for tubers and roots.

Fig. 2

To evaluate which vegetable food was available to early humans and what role fire played, we examined the edible wild plants that grow in the savannas of the *Cradle of Humankind* today as an example. Multiple paleoanthropological sites were documented in this area in the northeast of South Africa. Among other things, this region is known for the oldest remains of *Homo erectus* in southern Africa and the earliest evidence of human interaction with fire from Wonderwerk Cave, dated to one million years before present. Many of the edible wild plants available today were probably also available more than two million years ago and had the same nutritional properties as today. Plant species that are widespread in these savannas and could therefore represent an important source of food were of particular interest. We researched the edibility and known methods of preparation for the individual plant parts of the most important species. Only those parts of the plant were considered that were edible in large quantities and that could have been a relevant source of food for early humans. The edible parts of plants are very diverse and were grouped into five categories: Fruits, seeds, underground storage organs (roots, tubers, rhizomes), leaves (including stems, sprouts, flowers, buds), and "others" (here: sap, bark, wood).

1 Fruit from the baobab or monkey-bread tree *(Adansonia digitata).*

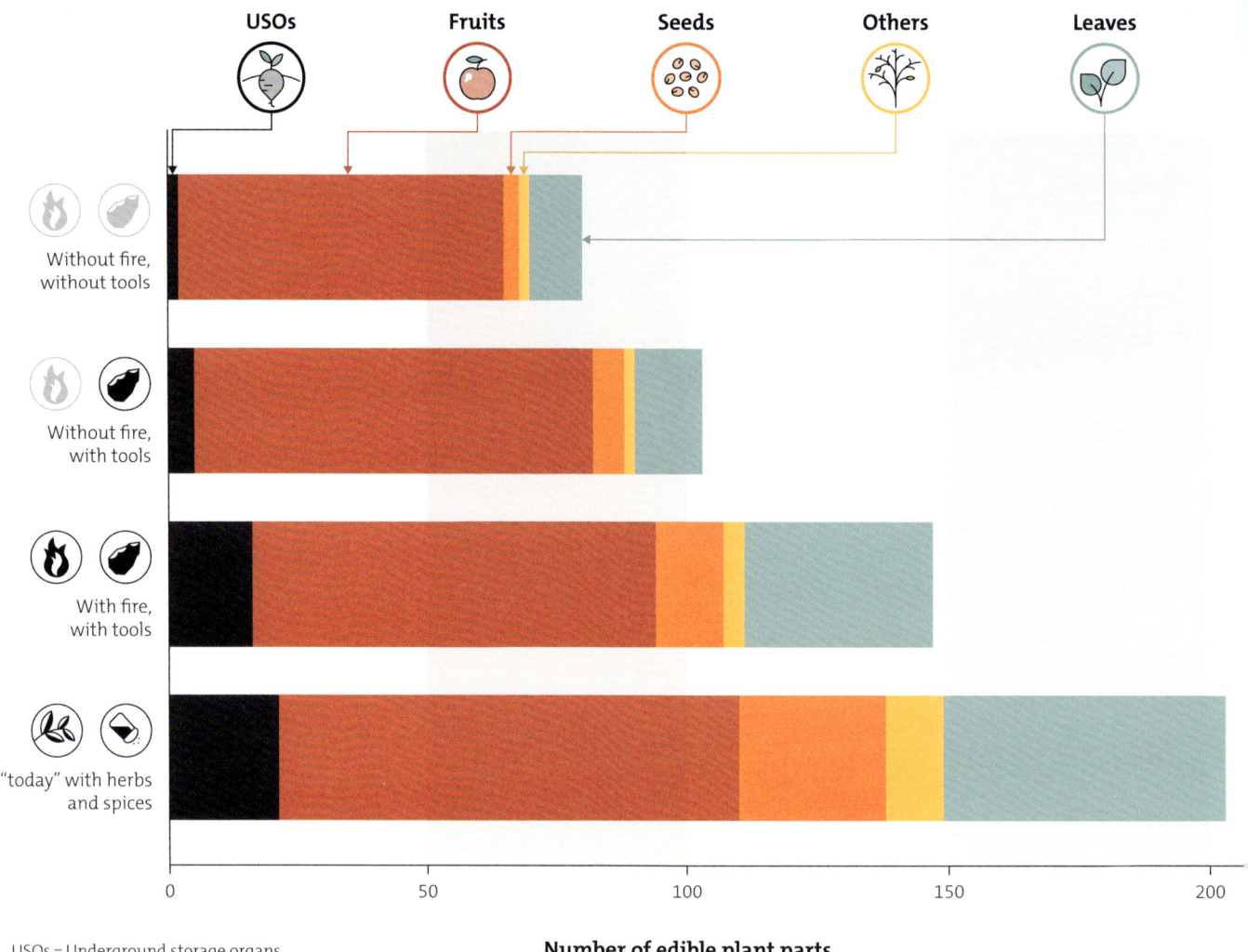

USOs = Underground storage organs

Number of edible plant parts

2 Number of edible plant parts available in the savanna in the *Cradle of Humankind*, South Africa depending on technological capabilities.

Fig. 2

Overall, there is an astonishingly wide range of plant-based food available in the South African savannas. Of 814 plant species, over 20 percent are described as edible, and many of these have multiple edible plant parts (see infobox on the baobab tree). More detailed information is available for 139 species, including 203 plants parts: even without any preparation, far more than half, namely 80 different parts of 74 plant species, are suitable for nutritional purposes. These include mainly fruits (63) and some leaves (10), which are available almost year-round in the mild South African climate. Many of the fruits are rich in vitamins and have a sweet, sugary pulp. They also have proteins and a higher fat content and are an important source of nutrition overall. With the help of stones or other tools, the menu can be extended to include over one hundred plant parts by the simplest means of preparation. This includes the removal of harder parts of plants by cracking as well as scraping and rubbing, which mainly accesses the starchy seeds and roots. Around half of all edible species documented in the region today can be prepared using these simple methods.

With the ability to control fire and use it to prepare food, the potential menu of humans expanded immensely. The number of edible plant parts now almost doubles, although here only plant parts are considered that can be made edible by simple cooking or roasting. Many starchy seeds and underground storage organs only become easily digestible through cooking or roasting and are therefore an important energy-rich source of food. This explains why, in the diagram, the proportion of edible underground storage organs tripled and that of seeds doubled with the availability of fire. Starchy storage organs in combination with fire, in particular, are assigned a significant role as a source of energy, because they also have the advantage of being available throughout the year, i.e., even during dry periods. The tubers of the wild potato *(Plectranthus esculentus)*, for example, can be eaten raw, roasted, or cooked and are very popular with the local population in South Africa today. But also the number of usable leaves and other plant parts that can be eaten as vegetables triples as soon as a fire is available as a form of food preparation.

All in all, mastering fire and using it for food preparation brought great advantages for nutrition—both in terms of the variety of edible plant parts as well as in terms of their exploitability. Many foods that can be eaten raw are easier to chew and easier to digest when cooked, and the nutrients they contain are better usable by the body.

Today's spectrum of species that are available for nutritional purposes is many times greater because the development of more complex processing and cooking techniques has resulted in the usability of additional plant species and parts. Herbs and seeds play a role as spices. Roots and tubers, whose inedible bitter substances must be removed through more complex processing steps, enrich the menu. Numerous wild plants are also used in the production of all kinds of beverages—from herbal teas and juices to beer, wine, and gin, and, last but not least, Amarula, a liqueur made from the tasty marula fruit of a wild tree *(Sclerocarya birrea)*. It is uncertain when our ancestors began to use spices and the like. Even if the exact timing of this development from the first mastery of fire to simple cooking and roasting to complex cooking is far from clear, our results show that the use of fire for food preparation, in particular, is a decisive step towards effective food yield.

Fig. 3

The baobab
or monkey-bread tree
(Adansonia digitata)

This imposing tree, which is widespread in the African savanna, provides the most edible plant parts. Even without the use of fire, five of them are readily available to eat. The fruits of the baobab tree *(Fig. 1 and below)* contain a floury pulp that is very nutritious and rich in vitamin C, as well as fatty, nutrient-rich seeds that are very tasty and easy to digest raw and roasted. Blossoms, young shoots, and root shoots are also edible. The leaves can also be eaten as vegetables when cooked over a fire. They are high in protein and are often used as an ingredient in sauces. Both the fruits and the dried leaves can be stored for several months. The soft, fibrous wood of the baobab tree stores a lot of water and is also a source of water. The fruit from the baobab tree is a highly valued part of the diet of today's hunting and gathering groups in the Kalahari. One can assume that they were an important source of food for early humans as well.

3 Examples of edible parts of South African plants: **1** sugary fruit of a type of fig *(Ficus sur)*, **2** tasty fruits of the bird gooseberry *(Hoslundia opposita)*, **3** *Ceropegia barberae*, whose root tuber can be eaten raw, **4** wild potato tubers *(Plectranthus esculentus)*, a popular source of starch, **5** *Phoenix reclinata* whose juice is used to make palm wine, **6** fruits from the marula tree *(Sclerocarya birrea)*.

Further reading

Berna, F./Goldberg, P./Horwitz, L. K./Brink, J./Holt, S./Bamford, M./Chazan, M. 2012 Microstratigraphic evidence of in situ fire in the Acheulean strata of Wonderwerk Cave, Northern Cape province, South Africa. Proceedings of the National Academy of Sciences 109, 2012, E1215.

Hardy, K. 2018 Plant use in the Lower and Middle Palaeolithic: food, medicine, and raw materials. Quaternary Science Reviews 191, 2018, 393–405.

Hardy, K./Brand-Miller, J./Brown, K. D./Thomas, M. G./Copeland, L. 2015 The importance of dietary carbohydrate on human evolution. The Quarterly Review of Biology 90 (3), 2015, 251–268.

Henry, A. G./Büdel, T./Bazin, P.-L. 2018 Towards an understanding of the costs of fire. Quaternary International 493 (10), 2018, 96–105.

Herries, A. I. R./Martin, J. M./Leece, A. B./Adams, J. W./Boschian, G./Joannes-Boyau, R./Edwards, T. R./Mallett, T./Massey, J./Murszewski, A./Neubauer, S./Pickering, R./Strait, D. S./Armstrong, B. J./Baker, S./Caruana, M. V./Denham, T./Hellstrom, J./Moggi-Cecchi, J./Mokobane, S./Penzo-Kajewski, P./Rovinsky, D. S./Schwartz, G. T./Stammers, R. C./Wilson, C./Woodhead, J./Menter, C. 2020 Contemporaneity of *Australopithecus*, *Paranthropus*, and early *Homo erectus* in South Africa. Science 368, 2020, eaaw7293.

Marlowe, F. W./Berbesque, J. C. 2009 Tubers as fallback foods and their impact on Hadza hunter-gatherers. American Journal of Physical Anthropology 140, 2009, 751–758.

PlantZAfrica http://pza.sanbi.org

Roebroeks, W./Villa, P. 2011 On the earliest evidence for habitual use of fire in Europe. Proceedings of the National Academy of Sciences 108 (13), 2011, 5209–5214.

The Useful Tropical Plants Database http://tropical.theferns.info/

Homo habilis

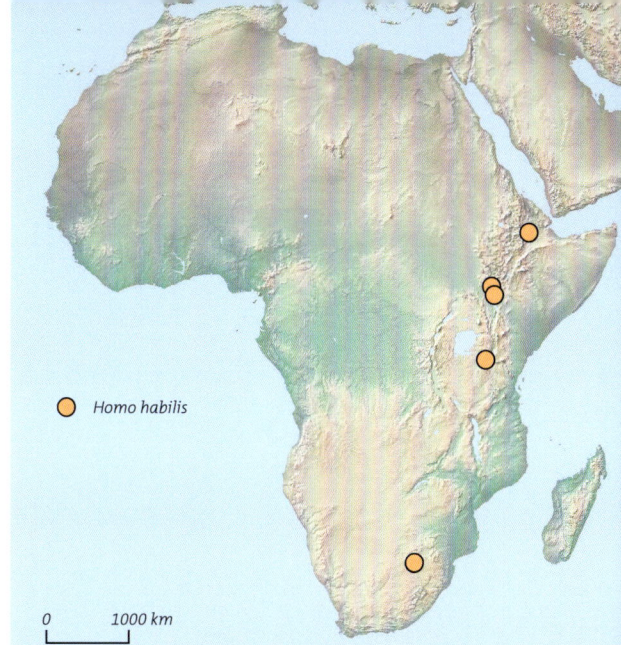

○ Homo habilis

0   1000 km

Discovery

The first *Homo habilis* find – a lower jaw bone – was discovered in 1959 by Heselo Mukuri at Olduvai Gorge.

Sites

Kenya: Koobi Fora, Ileret.

Tanzania: Olduvai.

Ethiopia: Hadar.

South Africa: Sterkfontein, Swartkrans, Kromdraai.

Finds

lower jaw bones, skull bones, teeth, hand and foot bones, upper arm and thigh bones.

Age

2.3–1.5 million years.

Brain size

590–687 cm³, possibly greater than 800 cm³.

Characteristics

Homo habilis is the most difficult species to grasp. For a long time, individual finds were assigned to this species because it was the only described human species with this old age. Fossils that belong to *Homo habilis* show features of both australopithecines, for example the shape of the teeth, and *Homo* species. They had remarkably small brains compared to other members of the genus *Homo*. *Homo habilis* was placed in the genus *Homo* due to the stone tools found nearby. In the 1960s it was assumed that only real humans, i.e. those of the species *Homo*, could make tools. Today we have evidence of stone tools from time periods long before *Homo habilis*. They were probably in use before the genus *Homo* evolved.

Facial reconstruction

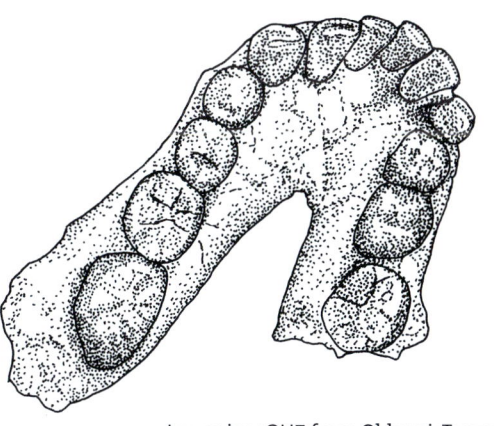

Lower jaw OH7 from Olduvai, Tanzania

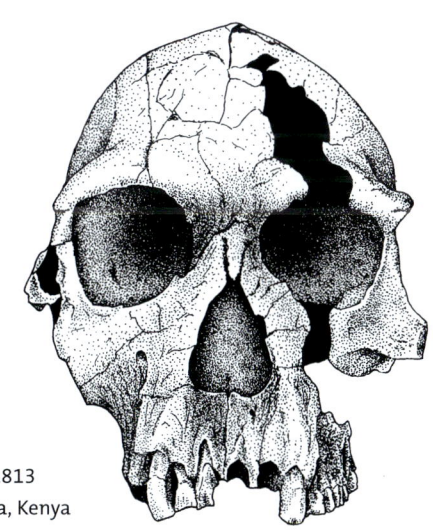

Skull KNM-ER 1813
from Koobi Fora, Kenya

Miriam Noël Haidle

Taking a detour on the path to human thinking

Thinking—not just a brain thing

We think with our heads, or more precisely with our brains, right? In our brain, sensory impressions are bundled, filtered, linked with one another, compared, and evaluated. New plans are forged and reactions postponed so that we can think things over. But does all this only take place in the brain? Is our brain a biological central computer, our thinking simply information processing?

Fig. 2

In addition to our brain, our whole body plays an important role in the way we think. Our senses supply us with impressions. As humans, we see three-dimensionally and in color, we hear particularly well in the frequency range of human speech, we perceive clear smells, and have a good sense of balance, which helps to precisely control our movements. With the proverbial "eagle eye", eagles see sharply at much greater distances and can also perceive ultraviolet light. Bats and dolphins orientate themselves and communicate with each other with the help of ultrasound. Dogs can sniff out much finer scents than we can. Birds have different organs of equilibrium for flight on the one hand and standing and walking on the other.

Thinking by acting

Our sensory organs are not simply suppliers of information. They are not static, any more than our nervous system and its particularly conspicuous part, the brain. They can change to a certain extent, depending on how they are used in the course of our lives. We learn. Babies explore their environment by touching as much as possible, experiencing cold, hot, wet, sharp, and cuddly things, by putting everything in their mouths and experiencing mixtures of sour, salty,

---

1 Aside from the brain, various senses and hand motor skills are also necessary to perceive, remember, and plan in the production of tools.

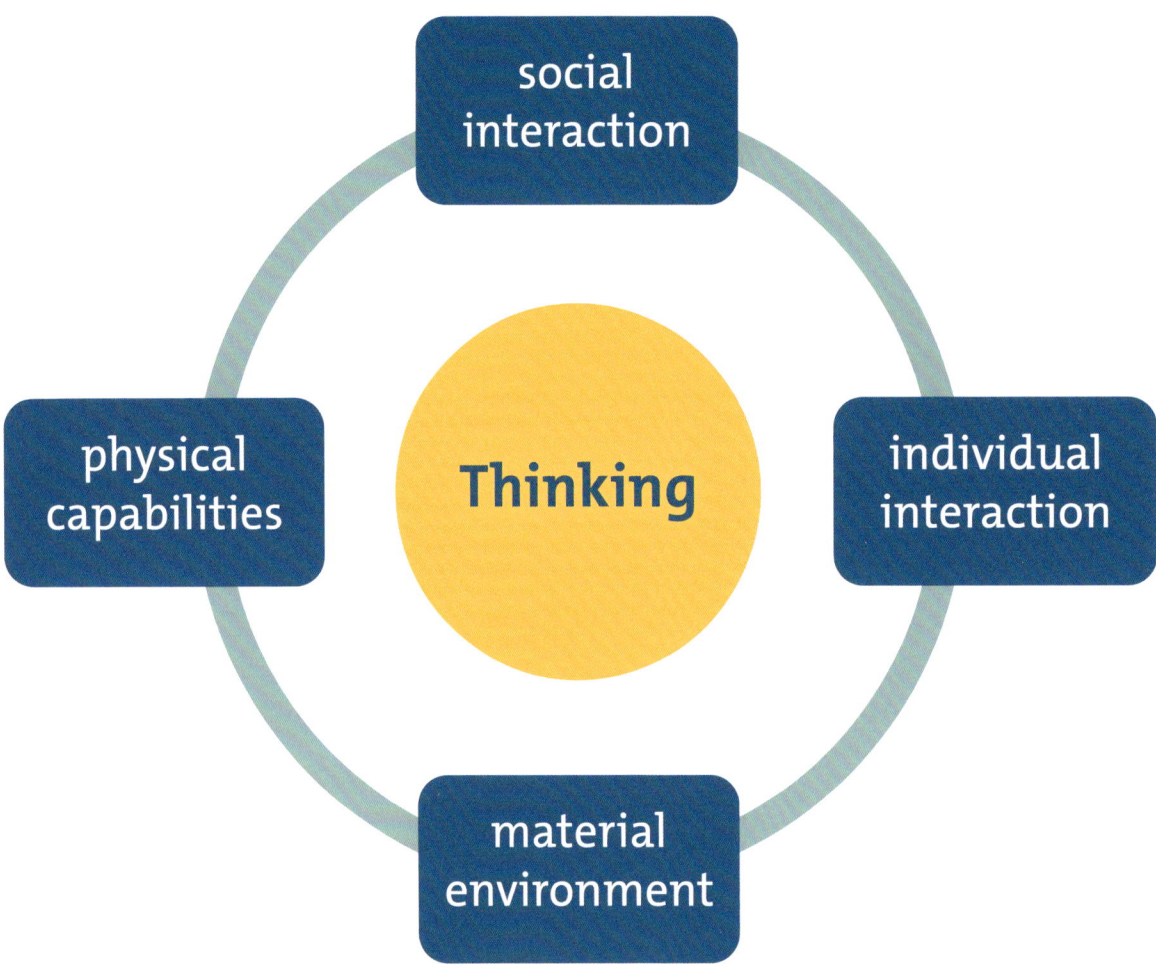

social
interaction

physical
capabilities

Thinking

individual
interaction

material
environment

2 The ability to think is influenced by many factors.

sweet, and dusty, by throwing things around, by pulling themselves up and falling down. We don't just see blue things as blue but learn to pay attention to certain sensory impressions and classify them as blue. Through our actions, we learn to control our bodies and to think: through different perceptions, the comparison of sensory stimuli with familiar ones, evaluations, resulting reactions and the generation of new perceptions.

Thinking embedded in the environment

Our thinking is embedded in how we deal with our environment. When a child learns to ride a bike, it won't do so if we tell it how it is done. It has to sit on the saddle, learn how to pedal, steer, brake, and stay upright all at the same time. It has to get a feeling for the vehicle. The muscles and senses develop routines so that the riders can focus their attention on special events ("a ball rolls into its path!"). It is said that once you have learned to ride a bike, you will not forget it. This applies to many things that have become a habit. In this case, our body is thinking, supported by the specific properties of things it interacts with. In addition, tools can support our way of thinking. A blind person can "see" their surroundings through a cane. A shopping list reminds me to buy the yogurt. In these cases, our perception and memory, i.e. parts of our thinking, are expanded by things outside of our body. And our way of thinking is only partly individual. As social beings, we benefit from the knowledge and experience of others, take on classifications and assessments, learn from others to pay attention to things, and think in certain ways. Our way of thinking is not trapped in our person but distributed within the group.

Development of thinking

In the course of a lifetime, the way we think develops and changes. We are continuously collecting knowledge, gaining experience, rearranging things accordingly, starting to be interested in something, pursuing a thought and expanding it, losing the thread, or completely rethinking something. We develop our way of thinking individually, each with his/her mix of personal and shared experiences, suggestions, and inhibitions. And we develop our way of thinking in an environment shaped by history and social contacts. Depending on the group to which I belong, I adopt different values, experiences, actions, and explanations. A single mother working in a shop who organizes her everyday life with a smartphone thinks differently than a farmer who lived during the Middle Ages and could neither write nor read. The mixture of prevailing traditions in attitudes, actions, and material environment defines our respective culture. In the course of the past three million years of human history, our physical prerequisites for thinking have changed alongside and with cultures. Human brains have grown from the size of a fist to the volume of a packet of milk and beyond. The density of nerve cells has increased, the relationship between different areas of the brain and their metabolism has changed. Our hands have become increasingly suitable for powerful grips on the one hand, and very precise handling of things on the other. Both hand-eye coordination and the fine motor skills of the hands have increased.

A growing ability to communicate finally culminated in many thousands of languages through which we can exchange ideas about the past and the future as well as about concrete things like cucumber salad or ideas like justice. Human thinking developed in the interplay of individual, historical-social, and evolutionary-biological developments with an environment increasingly shaped by humans.

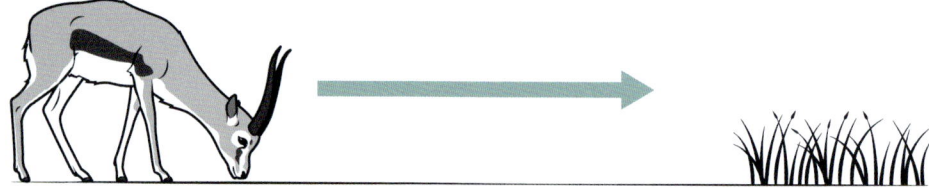

Immediate food intake
(without tools)

3

Simple use of tools
can be observed among multiple animal species.

4

3 When an antelope is hungry, she grazes grass—without detours.

4 If a chimpanzee wants to eat nuts surrounded by a hard shell, it must first find a suitable hammerstone to crack the nutshells. It must think about a detour before reaching its goal.

Development of human thought—the basics

The further we go back in time, the more difficult it becomes to figure out how and what people thought about. For periods without a written language, only the results of people's actions can provide us with clues about their ways of thinking. The production and use of tools offer a place to start. Different animals use tools and also make them. They use the tools for purposes that they could not achieve based on their physical abilities: the woodpecker finch pokes cactus spines into the wood to catch insects and maggots; capuchin monkeys crack hard nuts and clams with stones; orangutans use leaf cushions to protect themselves when climbing thorny trees. No species (except humans) is more proficient in handling tools than chimpanzees (see Wittig's article in this volume), who use them to get to hard-to-reach food, to draw liquids, to impress others, to cleanse and defend themselves. In terms of thinking, what makes tool behavior significant is the mental detour that an individual takes to reach its goal. While a hungry antelope only looks for grass and eats it as soon as it has found it, tool users must first look for a suitable device before they can start working on their actual object of desire. To do this, they have to put aside their actual goal and first focus their attention on the tool. The distance between problem (e.g. hunger) and solution (e.g. food) increases. Chimpanzees can use various tools to reach one goal.

5

Multicomponent toolsets composed of primary tools are used by chimpanzees
to collect ants or extract termites or honey.

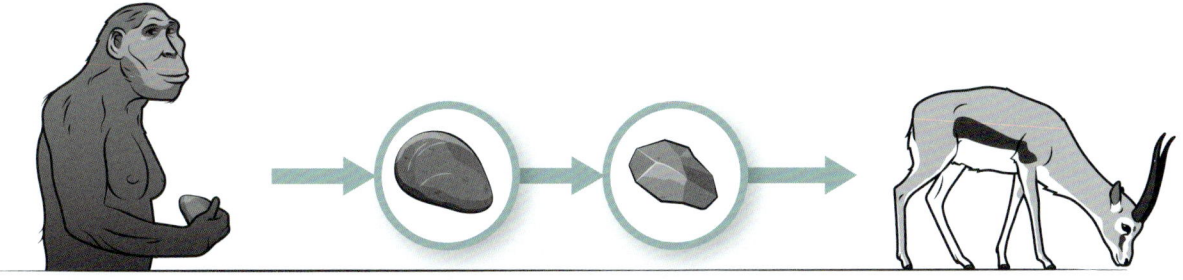

6

Using a tool to produce a tool
to reach a certain goal – since at least 3.3 million years ago.

Extended detour-thinking

The first evidence for an even greater detour in thought is around 3.3 million
years old. In Lomekwi, Kenya, stones were used to chop off sharp flakes from
other stones to work something with them. The use of tools to make tools to do
something else has only been documented for hominins. This trick of the again
expanded detour made it possible to create new tools with previously unknown
properties that opened up new possibilities. With cutting edges, for example,
humans could quickly cut off parts of a carcass or prepare wood for other tasks,
without having to use sharp teeth. The greater the detour, however, the more
thought and, in cases, planning is involved. To make a cutting stone flake, you
needed a good hammerstone and raw material suitable for knapping. If you were
lucky, both were close at hand when you wanted to steal a piece of the leopard's
prey. But if you first need to start an elaborate search for materials, food competitors like hyenas got there first. Those who could pay attention to several things at
the same time, remember places with good raw materials, and think ahead, had
to rely less on luck and had an advantage.

5 If a chimpanzee wants to treat
itself to a termite snack, he often has
to use two different tools: a stick to
break open the termite mound and a
thin twig to fish for termites.

6 The manufacture of stone tools
with a hammerstone, for example, to
dissect an antelope, requires extended thinking along with multiple
detours. So far, this is only known
for hominins.

Bit by bit

To be able to master further thought detours, it helped to break them down into small stages and arrive at the goal step by step. The time span between a need and its satisfaction became longer due to the detours. They diverged more and more until independent small units of action emerged. These fulfilled intermediate goals such as the procurement of raw materials or the production of tools—regardless of whether they were required immediately or not. These small units, called modules, had many advantages. Detached from an urgent need, the materials or tools could more easily be used for other purposes. They could be linked to one another in different ways. And broken down into small units, even more complex actions could be learned more easily.

Chimpanzee children take about three years to master the cracking of nuts with a stone. This is an indication of how long it may have taken hominins to learn the longer detours and to acquire the various skills and knowledge associated with them. By learning individual modules bit by bit, it was easier to acquire longer detours in thinking. With the help of flakes, children were able to practice cutting before they managed to make such devices themselves. They grew up in a group with the idea that stones could be shaped and knapped. Together with the elders, they would hike to places with good raw materials. They would learn about suitable stones when they helped carry the stones selected by their experienced elders back home. When the children were finally big enough to try their hand at stone knapping, they had already learned a lot about what it takes to produce and use the tools.

7 In the course of human history, humans not only strengthened their connection to members of their group on the one hand and improved upon the materials used on the other. The interaction of humans and objects became more and more intermingled, and the human ecological environment expanded.

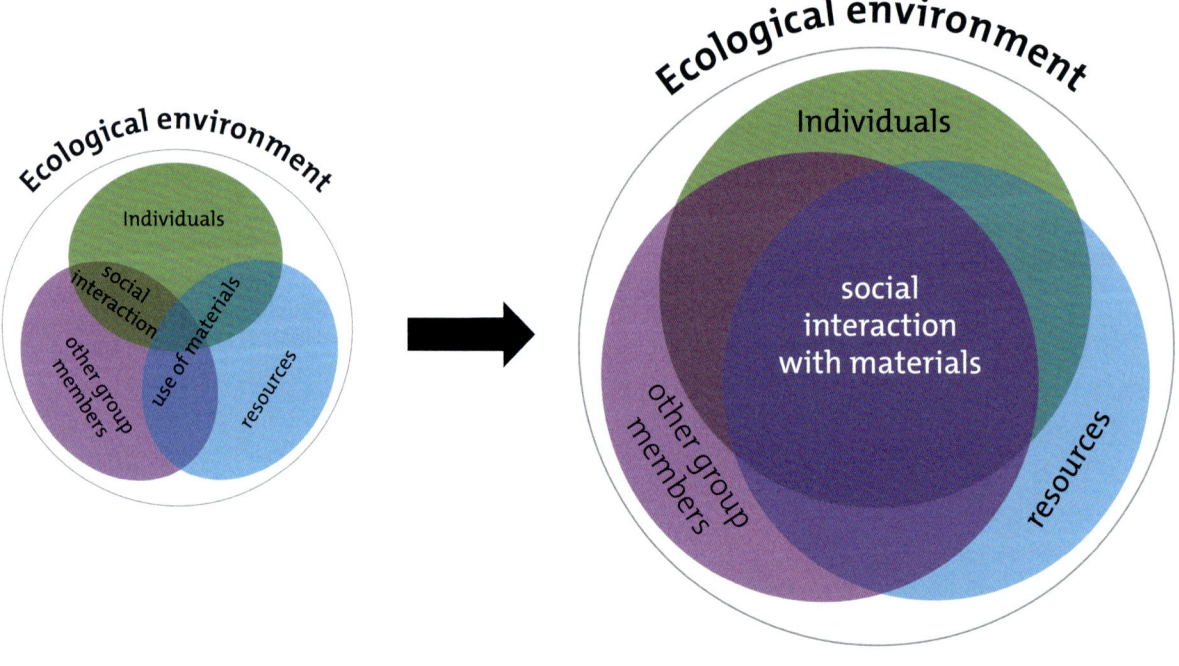

Together

If there is some space between a need and its satisfaction, if detours become conceivable, group members can also be more easily included in one's thoughts and actions. If I can break down my path from problem to solution into little modules and put them back together into large chains of action, then I can let other people take over individual parts. Different knowledge and different skills among different group members can be linked more easily. It is easier for everyone to benefit from the skills of others. In the beginning, hominins may have only used what others had left behind to continue with their actions: unused raw material, for example, or a reasonably sharp flake. Then several individuals might carry raw material to a production site, but one person was more successful at stone-knapping and allowed others to use the surplus tools. To contribute to a common solution to a problem, not just by chance, but intentionally, different people had to be able to focus their attention on one thing: I carry this for you and you make this out of it for me. Gradually, real cooperation became conceivable. It was increasingly possible to think together with others, and others could be included in plans. By extending detour thinking to group members, the social bonds within groups were strengthened.

Different—and therefore more

Monkeys, and especially great apes, build strong social relationships with the members of their group. They are full of ideas when dealing with their environment. Through special actions (for example, processing nettles before eating among gorillas) and the use of tools (for example, making leaf sponges to scoop up liquids among chimpanzees) they can make use of a wide range of resources. Compared to other animals, their ecological environment is large and varied. In the course of human developmental history, extended detour thinking increased the degree of interaction with group members as well as tools and resources, and the communal sharing of materials opened up further possibilities. The entire ecological environment of early humans expanded by thinking in further detours. Each new thing that was included in the thought process and actions could become a bridge to another.

Fig. 7

From simple stone tools to machines, from supporting sounds to entire novels, from the use of natural fire to electricity, clothing, agriculture, art, religion, school, and science ... Everything that we take for granted today has its origin in the fundamental extension in detour thinking that began around three million years ago. By taking detours, humans were able to meet their environment more flexibly, adapt to new conditions, and thus colonize regions far away from their African origins.

Further reading

Fingerhut, J./Hufendiek, R./Wild, M. 2013 Philosophie der Verkörperung. Grundlagen-
texte zu einer aktuellen Debatte (Berlin 2013).

Haidle, M. N. 2012 How to think tools? A comparison of cognitive aspects in tool behavior
of animals and during human evolution (Tübingen 2012).
http://hdl.handle.net/10900/49627

**Tomasello, M./Melis, A. P./Tennie, C./Wyman, E./Herrmann, E./Gilby, I. C./Hawkes, K./
Sterelny, K. 2012** Two key steps in the evolution of human cooperation: The interde-
pendence hypothesis. Current Anthropology 53(6), 2012, 673–692.

Varela, F. J./Thompson, E./Rosch, E. 2017 The embodied mind. Cognitive science and
human experience. Revised edition (Cambridge/Mass. 2017).

Homo ergaster

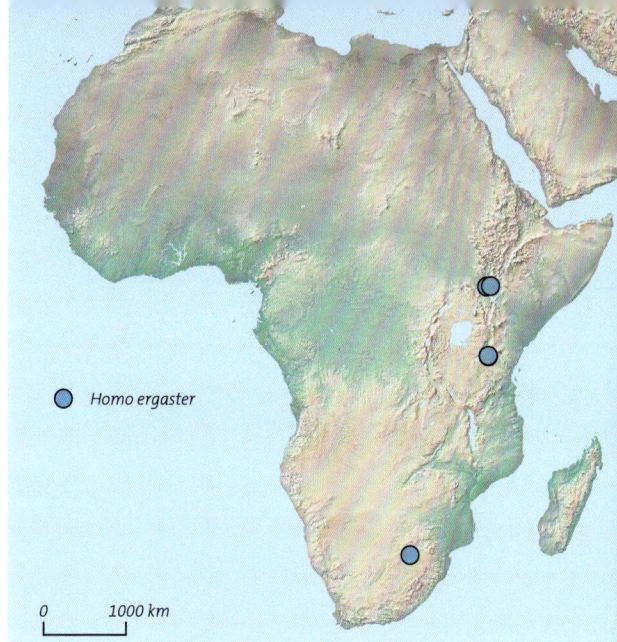

Discovery

Richard Leakey discovered the first remains of a *Homo ergaster* in 1971 in Koobi Fora, Kenya. Some of the fossils described by John Robinson in 1949 were subsequently assigned to the newly named species *Homo ergaster* in 1975.

Sites

Kenya: Koobi Fora, Lake Turkana.

South Africa: Swartkrans.

Finds

Skull with lower jaw bone, pelvic bones, shoulder bones, spinal column, arm and leg bones, skeleton of the "Turkana-Boy".

Age

1.9–1.4 million years.

Brain size

510–900 cm³.

Characteristics

Many researchers consider *Homo ergaster* to be the early African form of *Homo erectus*. The physical characteristics are very similar. In general, *Homo ergaster* were tall and graceful. They moved completely upright and were persistent runners. Based on the 1.6 million year old partial skeleton of the "Turkana Boy", which was 1.59 m tall, the size for adult individuals was calculated to be around 1.85 m. Although there is no clear evidence, tool use is assumed for *Homo ergaster*. However, in every complex in which stone tools were found in connection with these fossils, fossils of *Paranthropus boisei* are also found. They may also be considered as potential makers of the tools.

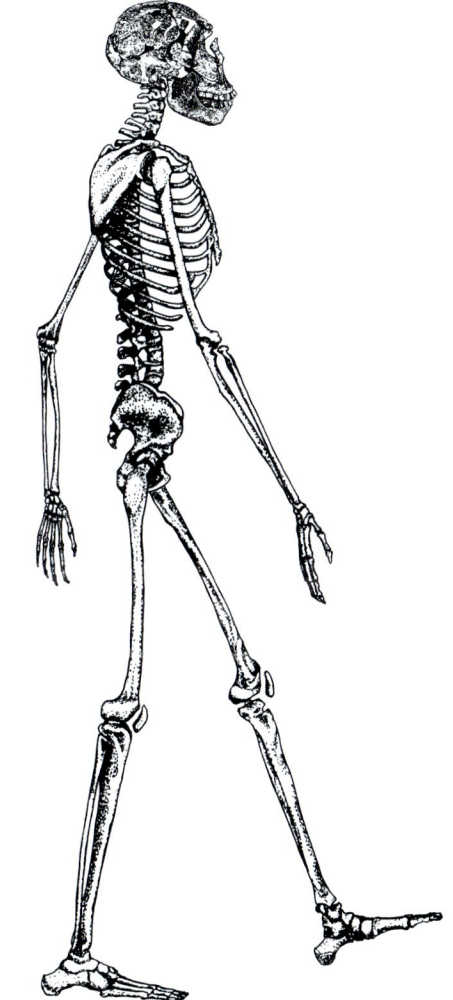

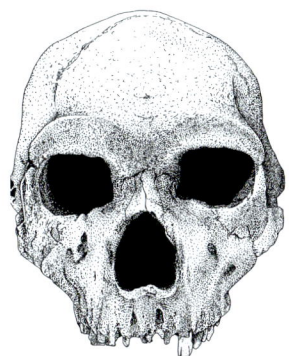

Skull KNM-ER 3733 from Koobi Fora, Kenya

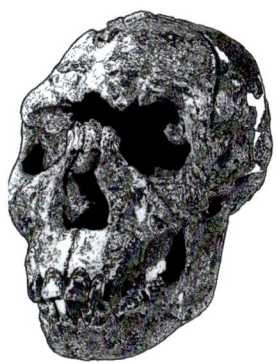

Skull of the Turkana Boy

Turkana Boy KNM-WT 15000 from Kenya

Christine Michel

Of rattles and puzzle boxes — social learning as the key to being human

Learning is great! It enables us to continuously develop and to create something new on this basis. Fortunately, not everyone has to reinvent the wheel. Instead, we build on the knowledge of our ancestors, constantly expand it, and collect it in encyclopedias or on online sites.

We all started small when we embarked on our learning expedition. It had already begun in the womb. An example: if fetuses are told a story repeatedly in the last month of pregnancy, they seem to remember it later. They will react to it specifically after they are born when they hear this story again. By the time we read this text, we have already left this early stage of development behind us and can no longer remember what it was like to see the light of day for the very first time. We can only imagine how exciting, loud, and perhaps also overwhelming it must have been to experience our world for the first time with all our senses. It is impressive how well newborn babies find their bearings. This raises the exciting question of how babies know which of the many pieces of information they are bombarded with is important and which they can simply ignore. Research in developmental psychology over the past few decades has a possible answer to this: the little ones follow the example set by their fellow human beings.

Newborn children find faces and biological movement, i.e., movement patterns that originate from living beings, particularly exciting. They are very good at recognizing other people's viewing direction and notice early on when they are being looked at. Where people look is an indication of what the person is paying attention to. For example, if a person keeps turning away from us during a conversation and looks to the side, we usually follow their viewing direction to find out what is so captivating.

1 A curious child.

2 The baby sits intently in front of the eye tracker, a special device that recognizes exactly where on the screen the child is looking.

Infants show similar behavior. They not only look to see what other people are looking at but also use other people's viewing direction to learn: A series of studies have shown that infants as early as four months can better process and recognize objects that another person is also looking at. How do we know? After all, we cannot ask the child. One way is to examine where children tend to look. You can film children and later evaluate the video accordingly. But there are also devices, so-called eye trackers, which provide computer-based information about what a person is looking at.

Fig. 2
In a series of studies, babies observed one person looking at a toy, such as a ball, and actively not looking at another toy, such as a rattle. The two toys were then shown again. On average, the babies looked longer at the toy that was not looked at by the person beforehand, i.e., the rattle in our example. What does this result imply? Infants prefer new things: they look longer at things that are new and therefore exciting for them. When babies see that another person is looking at the ball, the child's attention is drawn to it and the memory of the ball is stored. If the child sees the ball a second time, it is already familiar and therefore less interesting. The other toy, the rattle, is still unknown when it is

presented again and is therefore looked at for a longer time. From these results, researchers conclude that other people's viewing direction draws the babies' attention and helps them focus on important things in their environment.

But is that typically human behavior? We do not really know yet, but the assumption is that humans can draw a comparatively large amount of information from other people's viewing direction. Why is that? Compared to the eyes of other species, human eyes exhibit a particularly large contrast between the black pupil and the white background sclera.

Fig. 3

Could the typical human black and white contrast of the eyes help babies see what other people consider important? To find out, researchers showed different videos to four-month-old babies. In one video, the babies saw black dots that moved to the side on a white background, just like eyes that look to the side. The eyes "looked" in the direction of one toy and away from another. When the two toys were presented again, the babies—just as in the previous studies —looked longer at the toy from which the eyes had previously turned away. It seems that black dots on a white background have the same effect as a person's viewing direction: they draw the children's attention to things in the environment and the children learn something about them in a targeted manner. Later these things are more familiar and therefore less interesting and are only briefly looked at.

Fig. 4

In another experiment, the children saw the same video, only this time the contrast was reversed: white dots were moving on a black background. Now the "eye movement" did not have such a clear influence on how long the children looked at the two objects. That is, white dots on a black background guided the babies' attention less than black dots on a white background. These types of studies show us that others can steer the babies' attention. The eyes seem to play a key role.

Social learning, that is, learning from other people, is far more multifaceted. The older children get, the greater their scope for action and their motor skills. Children can now carry out complex actions in a targeted manner. A crucial ability is to observe others and to carry out seen actions oneself, i.e., to copy behavior. Puzzle boxes are often used in research to investigate the development of this observational learning.

You can perform various actions on such puzzle boxes, for example, pushing a lever, inserting a stick, or knocking on it. Most of these boxes contain a reward. In studies on observational learning, children are shown what to do to get the reward out of the box (for example, by sticking a stick in an opening). Then the children are allowed to try to solve the box themselves. Here it is examined whether the children achieve the goal and whether they imitate the actions that are necessary for it, that is, whether they have learned through observation.

Fig. 5

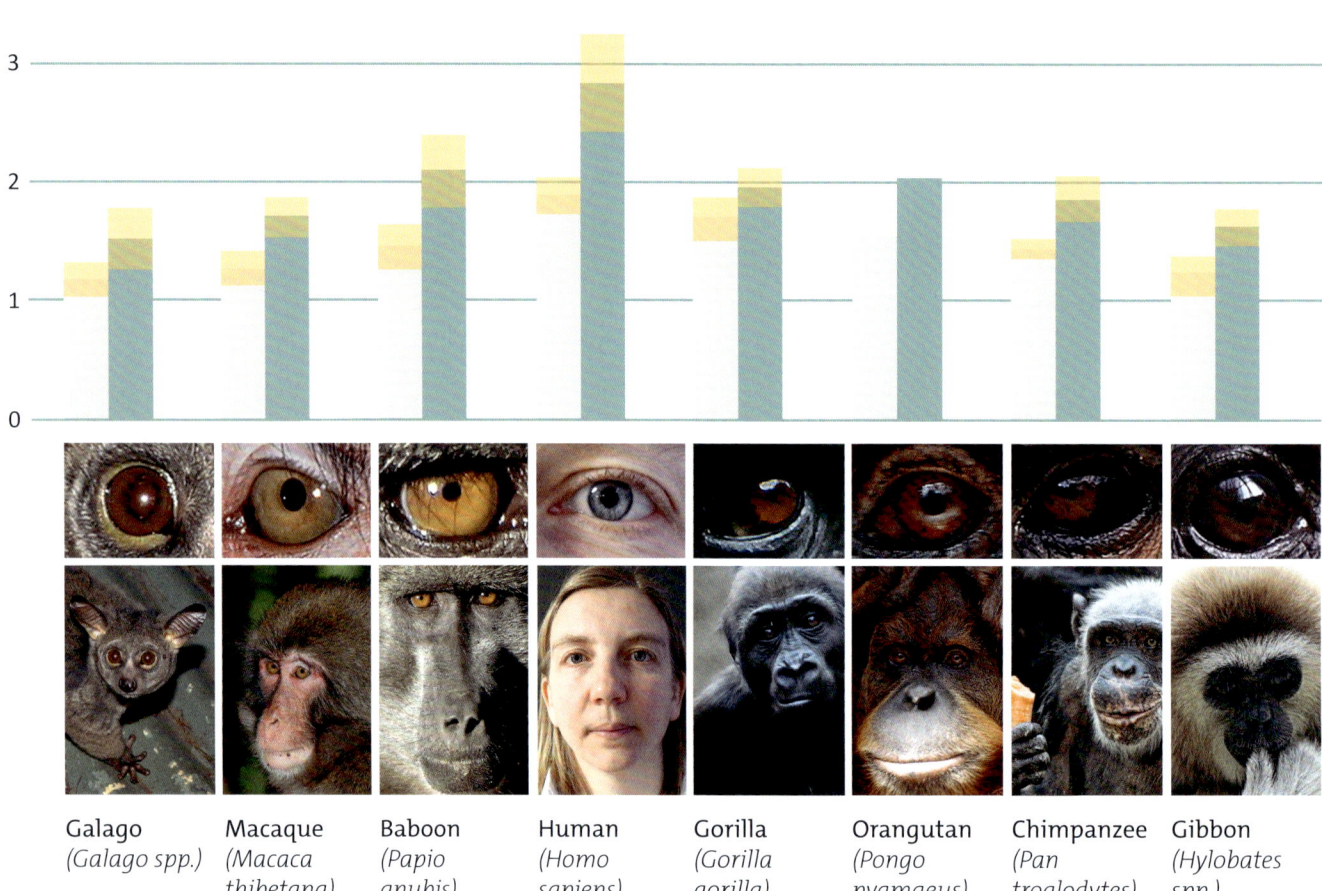

The ratio of the width of the white sclera and the colorful iris.

Width/height ratio of the contour of the eye.

Possible individual variation.

3 Comparison of the eyes of different species. The strong black and white contrast in human eyes, which helps people recognize in which direction their counterpart is looking, is especially striking.

In research, a distinction is made between two imitation strategies: so-called emulation and imitation. In emulation, a person learns something about the goal that can be achieved through observation. In our example, this means that you can get the reward out of the puzzle box. It is not important how exactly or by which means you get the reward. It is important that the reward can be taken out of the box. The learners could just as easily break the box, tip it over, or use another tool. Emulation processes, therefore, concentrate on the goal of an action, not on the way to get there. In contrast, in imitation, the journey is the goal. Here a person learns through observation how exactly an action must be

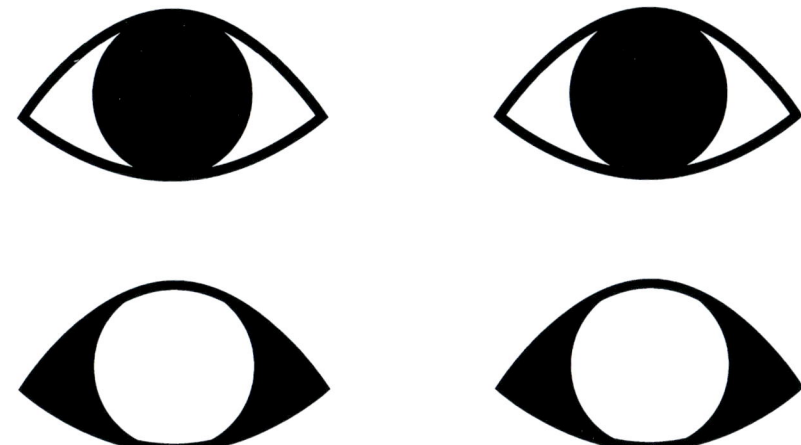

4 Illustration of the "eyes" that the babies saw in the study. Above, the eyes with natural black and white contrast. Below, the changed contrast with white dots on a black background.

carried out to achieve a goal. In the case of imitation, the learners would insert the stick in the same opening to receive the reward. With imitation, behavior is copied, with emulation the focus lies on the goal.

Do humans differ from other species in their imitation or emulation behavior? This question was investigated in a study with circa four-year-old children and chimpanzees. The study consisted of two different tasks: in one task, children and chimpanzees saw an opaque puzzle box with a reward inside. For the children, it was a sticker and for the chimpanzees, it was a treat. The investigators demonstrated different actions that could be done on the box. Some led to the goal: if you inserted the stick into the lower opening of the box, you could reach the reward. Another action, however, was pointless: if you inserted the stick into the upper opening, you could not reach the reward. Since the box was opaque, it was not clear why only the lower and not the upper opening led to the prize. The same procedure was also demonstrated using a transparent box. Here the observers saw that a built-in plate in the upper opening prevented them from reaching the reward. The mechanism of the puzzle box was obvious. Now it was time for the observers to try. Would children and chimpanzees imitate (that is, also copy the unnecessary actions and insert the stick in the upper opening) or emulate (that is, achieve the goal by only inserting the stick in the lower opening)?

Children imitated the actions on both puzzle boxes, that is, they also imitated the useless actions, regardless of whether they could see the mechanism in the box or not. Chimpanzees, on the other hand, only imitated the unnecessary actions if the box was opaque and they could not see the blocking plate. If the mechanism for releasing the prize was clear, they emulated and took the reward out of the lower opening. But why did the children imitate unnecessary actions and chimpanzees not? What does this result tell us about differences in social learning between the human and chimpanzee species?

5 Example of a puzzle box as it is used in the studies on observational learning.

Current research discusses a variety of explanations for why children mimic unnecessary actions. For example, social norms could be seen as a reason to imitate something that does not make sense ("That's the way it is done!"). It is also possible to assume that the person who demonstrated the senseless action has a specific intention ("There's got to be a good reason why he or she does something so strange. I should do it!"). Or, children want to belong, and imitation, i.e., doing it exactly the same way, could promote a feeling of belonging ("If I do it exactly as they did, then I belong!"). In the course of evolution, humans began to live together in growing groups, and cooperation and collaboration became more important. We can therefore assume that it is particularly important for humans to give a good impression and to be part of the group, as this makes it easier for us to find partners with whose help our coexistence and survival becomes easier. In the case of chimpanzees, this aspect could be less pronounced, so that they concentrated more on achieving the goal than on social processes—and therefore emulate. Some researchers believe that chimpanzees are not able to learn via imitation.

In social learning, children seem to place great emphasis on social norms, affiliations, and intentions. This is supported by another finding: children let others dissuade them from their strategy when solving a puzzle box. If they observe their peers, they often tend to adopt their strategies. Great apes, on the other hand, do not care when another ape has found a new solution strategy for the puzzle box—they tend to stick to their own strategy. In their behavior, children are much more influenced by other children than great apes are influenced by other great apes.

There is a lively debate in science about why and under what circumstances humans and other species imitate unnecessary actions—and thus also under what circumstances they can best learn from like-minded people. It is believed that imitation and learning about useless actions are important for us as a human species to be able to pass on cultural knowledge. Cultural knowledge includes, for example, ritual processes or customs that have no obvious physical function but are of great importance within a culture. Pure emulation (reaching the goal) could make this information more difficult to pass on to the next generation or even lead to it getting lost.

Whether it's rattles or puzzle boxes: the field of social learning is broad and with the help of developmental psychological studies we are getting a little closer to solving the riddle of what it means to be human. At the same time, however, we are still a long way from understanding everything. This is what makes learning about ourselves as a human species so exciting!

Further reading

Dunn, K./Reid, V. 2020 Prenatal cognition. In: The Encyclopedia of Child and Adolescent Development (Hoboken 2020) 1–9.
https://doi.org/doi:10.1002/9781119171492.wecad137

Haun, D. B. M./Rekers, Y./Tomasello, M. 2014 Children conform to the behavior of peers; other great apes stick with what they know. Psychological Science 25, 2014, 2160–2167. https://doi.org/10.1177/0956797614553235

Hoehl, S./Keupp, S./Schleihauf, H./McGuigan, N./Buttelmann, D./Whiten, A. 2019 'Over-imitation': a review and appraisal of a decade of research. Developmental Review 51, 2019, 90–108. https://doi.org/10.1016/j.dr.2018.12.002

Horner, V./Whiten, A. 2005 Causal knowledge and imitation/emulation switching in chimpanzees (*Pan troglodytes*) and children (*Homo sapiens*). Animal Cognition 8(3), 2005, 164–181. https://doi.org/10.1007/s10071-004-0239-6

Kobayashi, H./Kohshima, S. 1997 Unique morphology of the human eye. Nature 387, 1997, 767–768. https://doi.org/10.1038/42842

Michel, C./Wronski, C./Pauen, S./Daum, M. M./Hoehl, S. 2019 Infants' object processing is guided specifically by social cues. Neuropsychologia 126, 2019, 54–61. https://doi.org/10.1016/j.neuropsychologia.2017.05.022

Reid, V. M./Striano, T. 2005 Adult gaze influences infant attention and object processing: Implications for cognitive neuroscience. European Journal of Neuroscience, 21(6), 1763–1766, 2005. https://doi.org/10.1111/j.1460-9568.2005.03986.x

Schleihauf, H./Hoehl, S. 2020 A dual-process perspective on over-imitation. Developmental Review 55, 2020, 100896. https://doi.org/10.1016/j.dr.2020.100896

Simion, F./Di Giorgio, E./Leo, I./Bardi, L. 2011 The processing of social stimuli in early infancy: From faces to biological motion perception. Progress in Brain Research 189, 2011, 173–193. https://doi.org/10.1016/b978-0-444-53884-0.00024-5

Tennie, C./Call, J./Tomasello, M. 2012 Untrained chimpanzees (*Pan troglodytes schweinfurthii*) fail to imitate novel actions. PLOS ONE 7(8), 2012, e41548.

Homo erectus

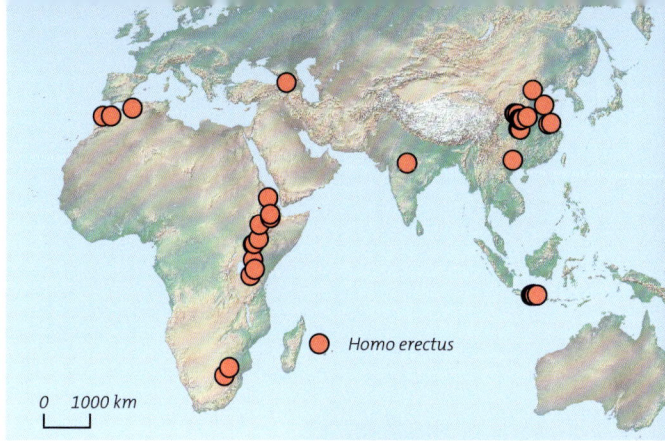

Homo erectus

0   1000 km

Discovery

The first remain of a *Homo erectus*, a cranium, was discovered in 1891 by Eugène Dubois in Trinil on the island of Java, Indonesia.

Sites

Indonesia: Sangiran, Sambungmacan, Trinil, Ngandong, Kedung Brubus, Mojokerto.

Georgia: Dmanisi.

South Africa: Saldanha.

Tunisia: Ternifine.

Further sites in: Kenya, Tanzania, Ethiopia, Morocco.

Finds

Skull fragments, teeth, lower jaw bones, various arm and leg bones.

Age

1.9 million–110,000 years.

Brain size

circa 870–1.150 cm^3.

Characteristics

In general, *Homo erectus* is considered to be the first species of the genus *Homo* to leave the African continent. However, since the remains from different regions differ greatly, it is not certain whether the finds from Africa and Europe are actually the same species as those of the Asian *Homo erectus*. *Homo erectus* were bipedal, just like modern humans today. Their size ranged from 1.45 to 1.80 m and they weighed between 50 to 60 kg. Their diet was presumably very variable and consisted of both plant and animal food. With the help of a particularly well-preserved skull, researchers found that *Homo erectus* already had cartilaginous noses, similar to those of modern humans, which led to improved thermoregulation of the breathable air and thus supported stamina and a more active lifestyle. In addition, they had a flexible thumb, which gave them fine motor skills. Skeletons of *Homo erectus* are very similar to those of modern humans and differ mainly in their stronger bone density and slightly elongated skulls with strong brow ridges above the eyes.

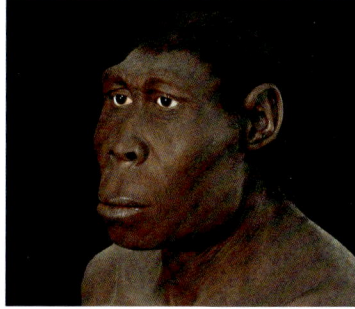

Facial reconstruction

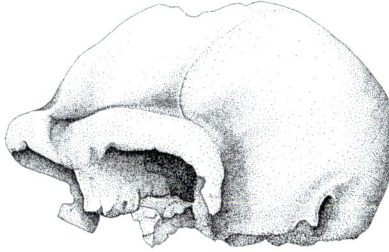

Skull calvaria OH9
from Olduvai, Tanzania

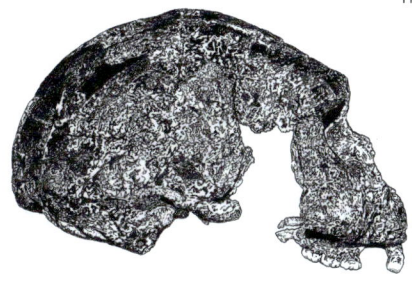

Skull
Sangiran 17, Indonesia

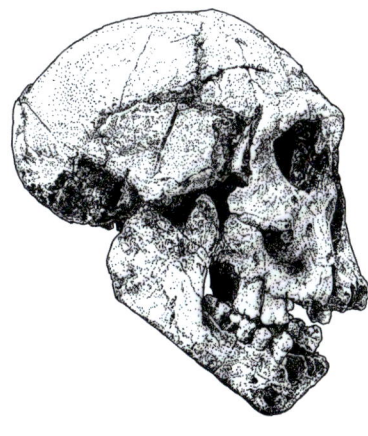

Skull D2700
from Dmanisi, Georgia

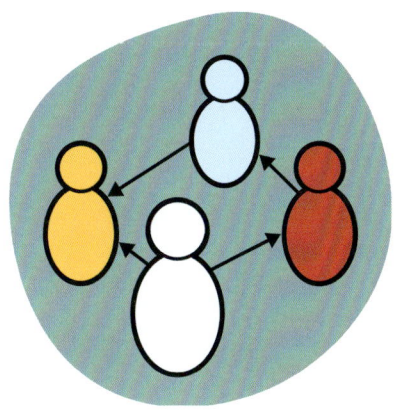

Oliver Schlaudt

Habitus:
the cultural primer

Four scenes

I

Susanne and Patrick enter the classroom. The air is stuffy and stale. Patrick grimaces: not a nice prospect of having to work here. Susanne opens the window without further ado, fresh air flows in. Patrick pauses. Why didn't he think of that himself? He knows how to open a window, he has done it a thousand times, and he also has his hands free. Still, he didn't do it, didn't even think about it.

II

Winter sets in early on the slopes of the Rocky Mountains. For the bighorn sheep, the food search becomes more difficult. Some animals will endure on the heights, but most of them will gradually drift into lower regions, fleeing the snow and cold winds. They follow a generation-old pattern. But they don't know it. The young animals follow the herd. In a few years, it will be they who initiate the annual migration.

Fig. 1

III

The group has made the descent into the narrow valley near the village of Langda in New Guinea. Some children are also part of the group. They look forward to romping around in the cool water while the adults look on the banks of the mountain stream for suitable stones for the manufacture of axes. The men quickly find one of the right size. They hold it, knock it with a large pebble. They are skeptical, someone shakes his head. The stone doesn't sound "right". They drop it and continue to work their way along the bank. The children have long since conquered the floods and are enjoying the cool reward.

Fig. 2

---

1 Young bighorn sheep in Badlands National Park, South Dakota, USA.

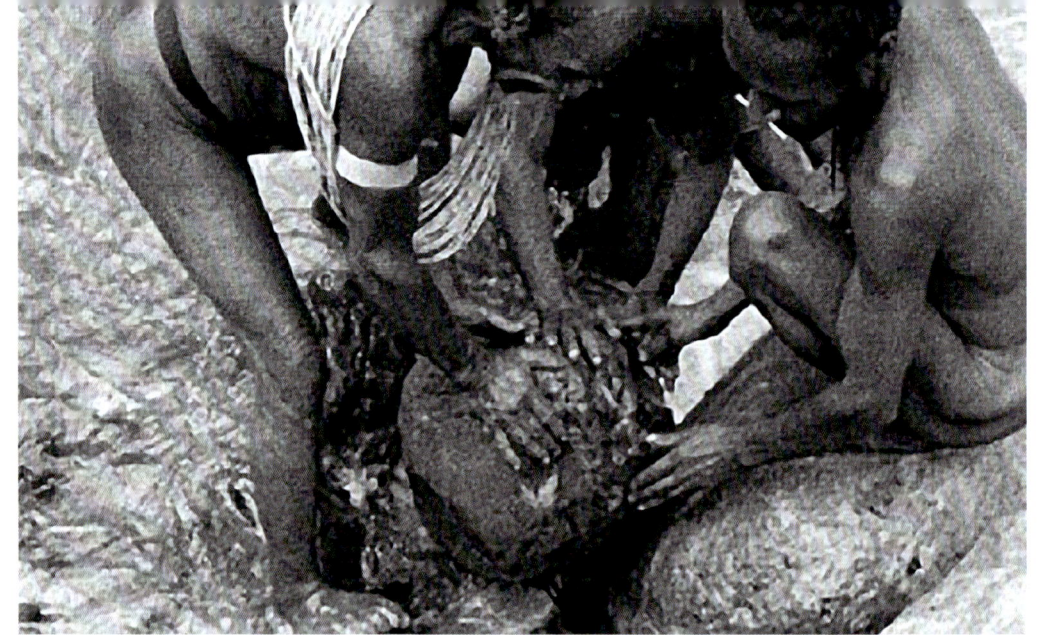

2 The search for and thorough examination of stones for the manufacture of ax heads in New Guinea.

IV

"Geopolitics", Armand answers frankly when asked about his interests. He is 17 and his face is still childlike. The chairman of the selection committee asks about the warring parties in the Syrian conflict. It is clear from his reaction that Armand does not know the answer. "Beginner's mistake", the chairman will say after Armand has left the room, and laughs slyly. "We'll take him". Hamid might have known the answer. The images of the civil war on the television fascinate him and he follows the news carefully when his brother doesn't zap to a different channel. But he doesn't like sitting here in front of the strange jury. He often looks down and reacts cautiously. He is denied access to the elite university.

Habitus

The French education system is a mystery to foreigners because striving for the elite and the demand for equality have gone hand-in-hand for two hundred years. The country, according to the consensus, needs a strong elite of engineers, military personnel, administrative officials, and diplomats. The only prerequisite for such a career is talent, regardless of whether you come from a Parisian dynasty or grew up on a farm in the provinces. Everyone should have the same opportunity at all times, provided the task corresponds to their capabilities. The fact that reality looks completely different, not only in France, is perhaps not surprising, but it is also not easy to explain. What does a child from a "good" background have, for example, that makes it successful in a job interview? Not the parents' money. There is also an "intangible" legacy. The sociologist Pierre Bourdieu called it "habitus" or "cultural capital." This is a milieu-specific attitude, but it is perceived as an individual trait at school or in an interview, for example. Susanne is more assertive than Patrick, Armand more confident than Hamid.

3 "Culture = capital", a light installation by Alfredo Jaar at the portico of the Braunschweig residential palace, 2016. Culture and capital have always had a tense relationship. The sociologist Pierre Bourdieu suggested that we view culture as a special kind of capital to explain the mode of operation and inheritance of inequalities in our society.

Pierre Bourdieu created the theory of cultural capital to understand how in-equalities are inherited and reproduced over generations in modern, egalitarian societies. This theory can also be useful to better understand the beginnings of human culture. The behavioral sciences and psychology have long empha-sized the importance of learning in the transmission of culture among many species of animals, especially humans. Cultural practices are passed on to the next generation not only through genes but also through cultural traditions. A distinction is made between different types of learning: mere imitation by the younger, conscious demonstration by the older, and finally the actual teaching, which is accompanied by corrective interventions and explanatory commentary. The habitus, on the other hand, refers to an even more fundamental way of learning. Neither the bighorn sheep nor the child of the Parisian upper class imitates an "action" previously observed from an elder. They do not experience the behavior of the elder as a planned, deliberate act in which a certain path is taken to achieve a purpose. They just join in, take the same path as previous generations, take part in conversations, develop a similar taste based on what their environment has to offer. The habitus is more like a basic mood in which the children bathe, which they soak up, and which they allow themselves to be carried along in. Bourdieu compared this to osmosis, the gradual seepage of a liquid through a fine-pored septum. At dinner, the children attend the conver-sations of their parents, learn what is important, how one talks about what, what one is at liberty to say, or what is forbidden. The children from New Guinea may not have paid much attention to what the older men were doing. Neverthe-less, they will later know unthinkingly how and where to find the suitable raw materials for the manufacture of the stone ax heads. Human culture grows out of this cultural humus.

Fig. 3

Further reading

Bourdieu, P./Passeron, J.-C. 1979 The inheritors: French students and their relation to culture. Transl. by R. Nice (Chicago 1979).

Haidle, M. N./Schlaudt, O. 2020 Where does cumulative culture begin? A plea for a sociologically informed perspective. Biological Theory 15, 2020, 161–174.
https://doi.org/10.1007/s13752-020-00351-w

Jesmer, B. R./Merkle, J. A./Goheen, J. R./Aikens, E. O./Beck, J. L./Courtemanch, A. B./ Hurley, M. A./McWhirter, D. E./Miyasaki, H. M./Monteith, K. L./Kauffman, M. J. 2018 Is ungulate migration culturally transmitted? Evidence of social learning from translocated animals. Science 361(6406), 2018, 1023–1025.
https://doi.org/10.1126/science.aat0985

Stout, D. 2002 Skill and cognition in stone tool production: an ethnographic case study from Irian Jaya. Current Anthropology 43(5), 2002, 693–722.
https://doi.org/10.1086/342638

Homo heidelbergensis

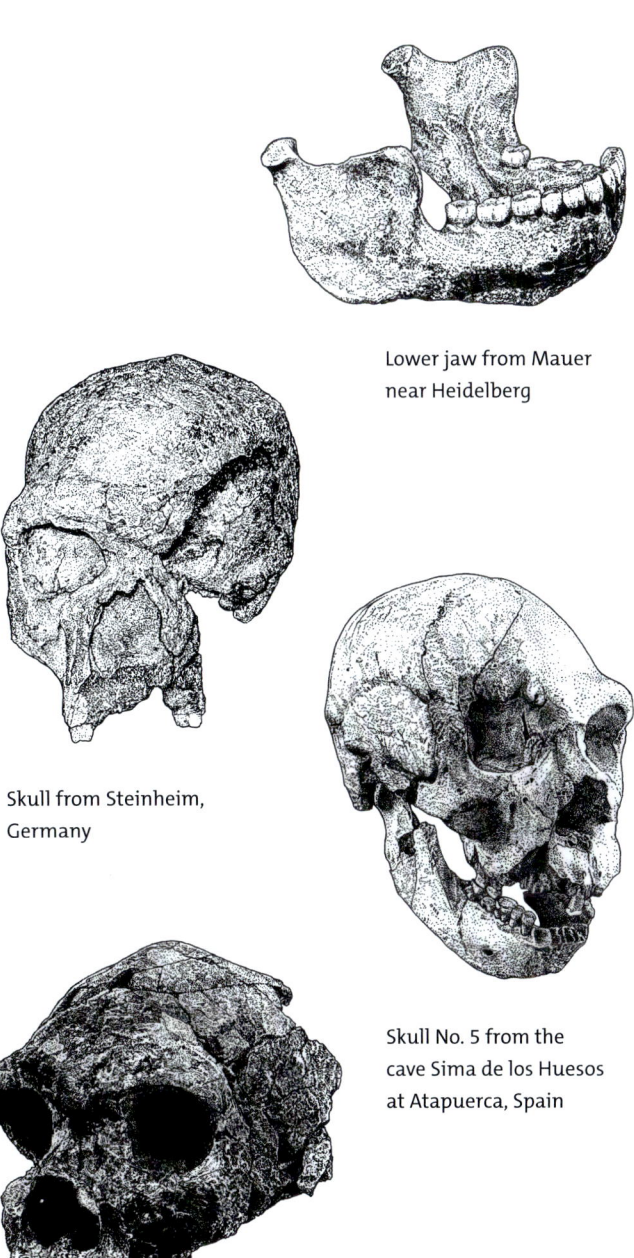

Homo heidelbergensis

0   1000 km

Discovery

A lower jaw bone discovered by Daniel Hartmann in Mauer near Heidelberg in 1907 is the first fossil of its kind.

Sites

Germany: Mauer, Steinheim.

Spain: Atapuerca.

Great Britain: Swanscombe.

France: Arago.

Hungary: Vértesszöllös.

Greece: Petralona.

Morocco: Thomas Quarry.

Israel: Zuttiyeh.

Ethiopia: Bodo.

Zambia: Kabwe.

South Africa: Elandsfontein.

Finds

Skull bones, lower jaw bone, arm and leg bones.

Age

600,000–200,000 years.

Brain size

circa 1,116–1,450 cm^3.

Characteristics

Homo heidelbergensis differ anatomically only slightly from *Homo erectus* and *Homo sapiens neanderthalensis,* which is why researchers to this day are discussing whether this is actually a separate species. Some features on the jaw and teeth speak for this, while the anatomical similarity to the Neanderthals and the similarly large brain volume speak against it. Genetic evidence also suggests a close relationship to the Denisova people. It is unclear to what extent African finds from the same period can be assigned to *Homo heidelbergensis*.

Like the Neanderthals, *Homo heidelbergensis* produced a variety of tools. Stone tools and the famous wooden spears and throwing sticks from Schöningen are also assigned to *Homo heidelbergensis*. The diet of the approximately 1.60–1.75 m tall and 60–80 kg heavy *Homo heidelbergensis* was presumably based – as in all hunter-gatherer societies – on a high proportion of plants.

Lower jaw from Mauer near Heidelberg

Skull from Steinheim, Germany

Skull No. 5 from the cave Sima de los Huesos at Atapuerca, Spain

Skull from Bodo, Ethiopia

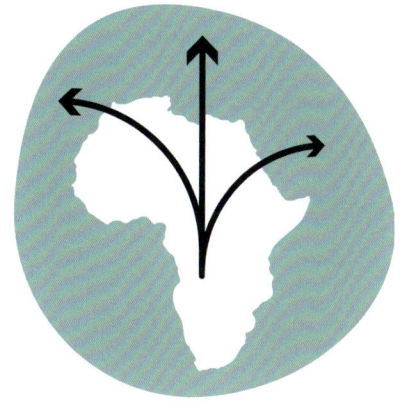

Miriam Noël Haidle

Across the mountains, into the wide world. Evidence of human expansion

Origin and expansion ideas

Just like the history of human cultural behavior, the associated history of human expansion across the world is subject to major changes. On the one hand, there have been many changes throughout the course of history: neither the cultural capabilities and their expression nor the course and speed of diffusion have remained the same. On the other hand, new discoveries and different approaches have shifted the perspective on the processes, so that the stories themselves had to be rewritten.

Charles Darwin's theories about the evolution of species and the recognition of the first fossil human remains in the Neander Valley near Düsseldorf about 150 years ago first allowed people to think about the possibility of pre-forms of modern humans from pre-biblical times. At the beginning of the 20ᵗʰ century, different forms from Asia *(Homo erectus)* and Europe (Neanderthals and *Homo heidelbergensis*) were known. However, when Raymond Dart presented the australopithecine child from Taung in 1925, the first hominin find from Africa as previously predicted by Darwin, no one wanted to believe in it. The idea of an early African origin of humankind was only accepted after the discovery of numerous additional finds from South and East Africa in the 1960s and 1970s. It was assumed that, following a worldwide dispersal, regional preforms evolved into *Homo sapiens*, the modern humans living today. In the 1980s, Africa came into focus as the singular region of origin of modern humans. Based on a new synopsis of the existing fossils, Günther Bräuer came to the same conclusion as Rebecca Cann and her colleagues based on initial genetic investigations: the common ancestors of our species, the genetic Eve, lived in Africa 200,000 years ago. In a second major expansion, Out-of-Africa II, between 60,000–40,000 years

1 Around 3.5 million years ago, humans were walking upright:
The cast of the Laetoli footprints clearly shows that the big toe was already aligned along the axis of the foot, as it is among humans today.

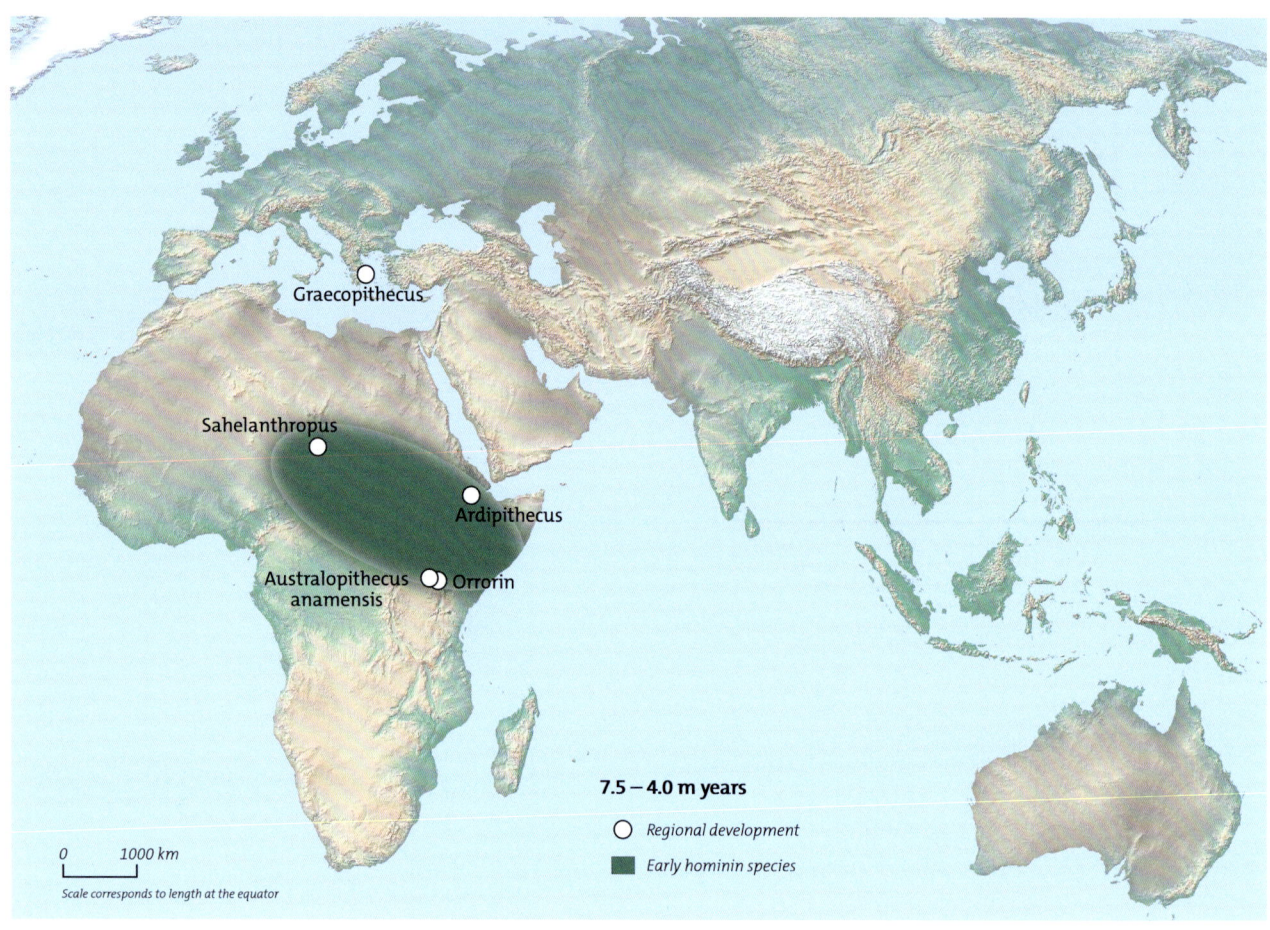

Graecopithecus

Sahelanthropus

Ardipithecus

Australopithecus
anamensis

Orrorin

7.5 – 4.0 m years

○ *Regional development*

■ *Early hominin species*

0   1000 km

Scale corresponds to length at the equator

2 Various forms of bipedal hominins are documented primarily from East Africa between 7.5 and 4 million years ago.

ago, anatomically modern humans displaced the original Neanderthal populations in Europe and *Homo erectus* in Asia. The first genetic studies on Neanderthals in the 1990s and 2000s seemed to confirm this picture.

Since 2010 our perception has changed yet again. Numerous fossil finds of known and newly discovered human species, improved dating methods, and more detailed genetic examinations of fossils indicate diverse expansion movements, regional origins of species, but also mixture of different human forms. Our species' past is much more complicated than previously believed.

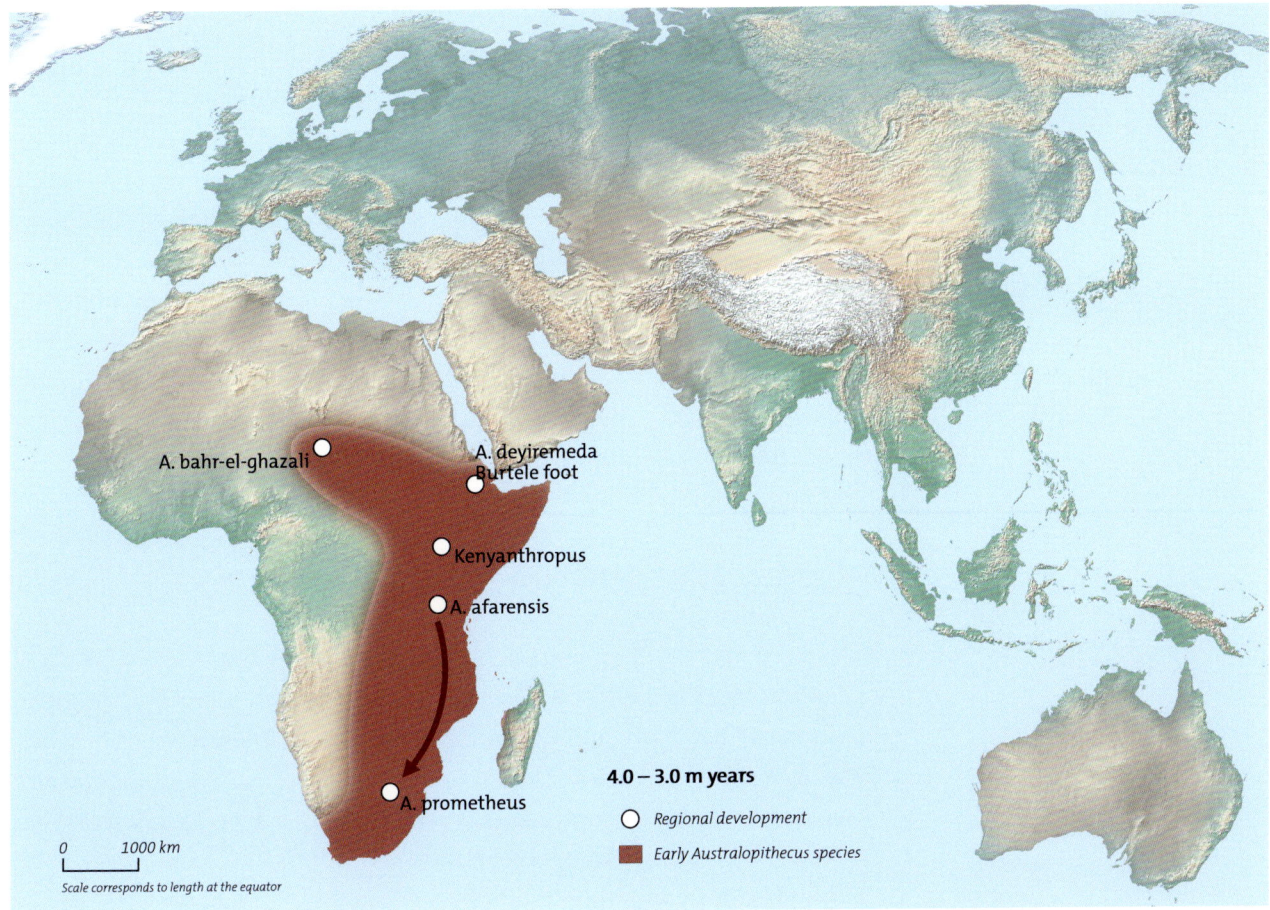

A. bahr-el-ghazali

A. deyiremeda
Burtele foot

Kenyanthropus

A. afarensis

A. prometheus

4.0 – 3.0 m years

○ *Regional development*

▨ *Early Australopithecus species*

0   1000 km

Scale corresponds to length at the equator

Beyond the known

Dispersal is not just dispersal. The geographic distribution of a species can change due to spatial shifts of the habitat. If the savanna grassland expands and pushes the forest back, savanna dwellers can spread out while forest dwellers retreat. Another possibility is the expansion into new habitats beyond the usual living conditions. This is not exclusive to humans but is a typical feature of the later human expansions. Targeted migrations, such as waves of emigration from Europe to America in the 19th century, are a relatively recent human phenomenon that is associated with the idea of a geographic destination.

As far as we can grasp them today, the early expansions of hominins and humans are not clearly defined events, but rather long-term processes, the exact courses of which are difficult to reconstruct. Early hominins inhabited a diverse habitat as early as 6 to 7 million years ago with patches of forests, palm groves, and grassy areas. A large part of their life took place on the ground and less and less in the trees. In the area of hominin distribution, a long-term trend towards

3 The australopithecines developed and spread to South Africa between 4 and 3 million years ago.

Fig. 2 a more open form of vegetation took place. The habitats became increasingly variable and localized with more as well as less dense tree cover, bush vegetation, and grasslands. We have too few finds from this early period to be able to retrace the expansions. But the early hominins already had properties that allowed them to cope with these environmental changes and thus to adapt to life in the changing habitats. *Graecopithecus freybergi, Sahelanthropus tchadensis, Orrorin tugenensis,* and *Ardipithecus kadabba* probably were not part of a single evolutionary line, but rather represented different parallel variations of bipedal locomotion. The function of the hands gradually changed in two ways: first, their role in locomotion, and thus their support function decreased. Secondly, improved fine motor skills in manipulating objects became more important— e.g., for the handling of tools and their manufacture.

Various adaptations and a corridor to the south

Between 4 and 3 million years ago the climate continued to cool and seasonal differences increased. East Africa became increasingly dry, and the grasslands expanded. During this period, the genus *Australopithecus*, in which bipedalism generally prevailed, developed in East and Central Africa. Different behaviors and food preferences made it possible for different forms to survive in one habitat. The first *Australopithecus* finds from South Africa are dated up to 3.67 million years ago. It is assumed that the australopithecines were able to spread south from East Africa due to their increased flexibility.

Fig. 3 Evidence from 3.3 million years ago points to a cultural innovation that brought about an expansion of the use of resources and shaped the whole of human history that followed. By using stone tools with sharp edges hominins were able to obtain parts of animal carcasses more easily and work on plants in different ways. It became easier to make use of different food options and thus to try out new methods. With the new technologies, the relationships with fellow hominins as sources of knowledge and experience also became more important. Both were beneficial as the environment continued to become drier and the variability of climatic conditions increased again around 2.8 million years ago. The continuous development of the East African rift system, with a lowering of the rift in a southerly direction and an uplifting of the rift shoulders, led to strong climatic differences between different regions. At times corridors opened or barriers formed, impacting the migration of animals and hominins between East and South Africa.

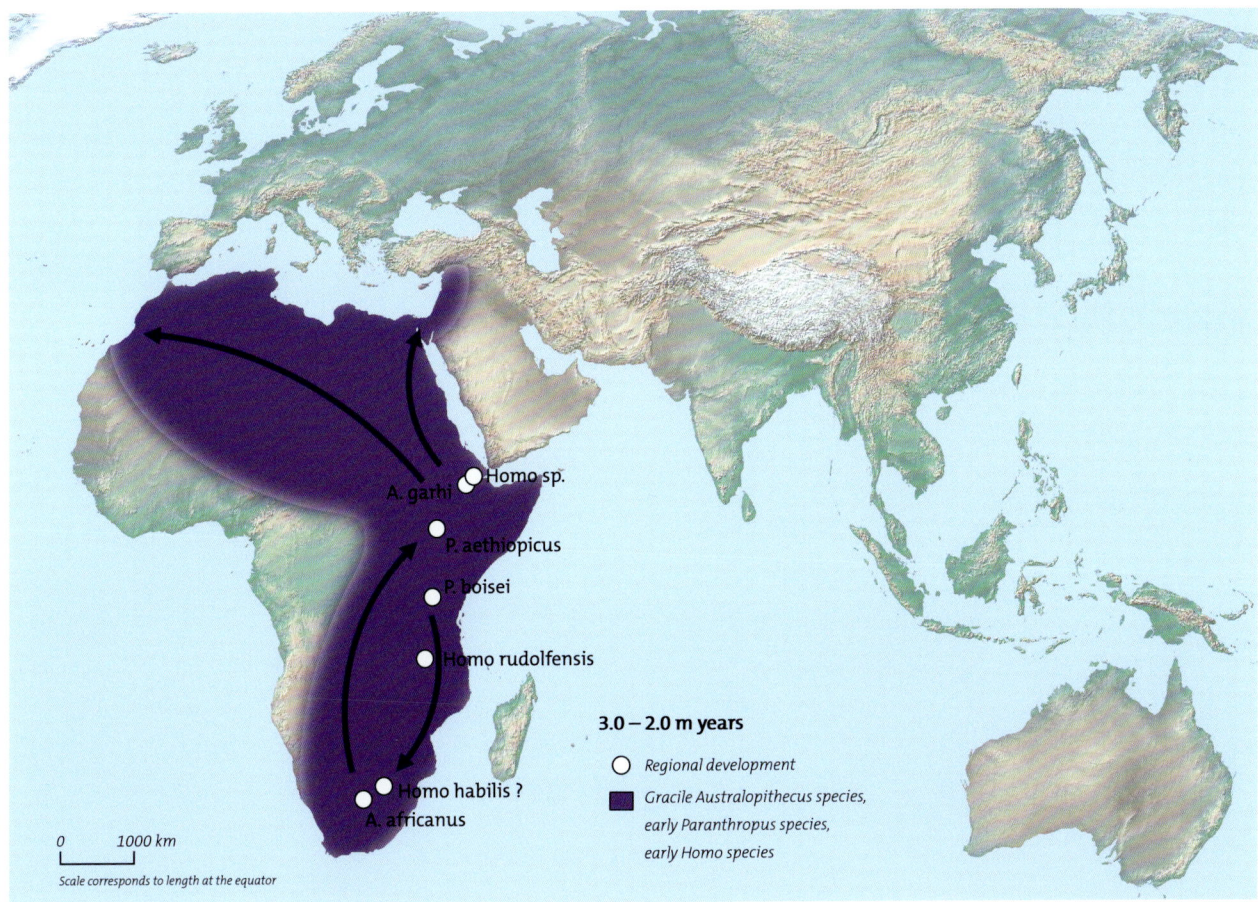

Fig. 4

3.0 – 2.0 m years

○ Regional development

■ Gracile Australopithecus species,
early Paranthropus species,
early Homo species

0   1000 km

Scale corresponds to length at the equator

A. garhi   Homo sp.

P. aethiopicus

P. boisei

Homo rudolfensis

Homo habilis ?
A. africanus

Into uninhabited lands

In addition to the australopithecines, two new genera developed between 3 and 2 million years ago. While *Paranthropus* included robust specialists for chew-intensive plant foods such as grasses and sedges, *Homo* increasingly specialized in "non-specialization" with smaller teeth, flatter faces, a slightly enlarged brain, — and above all, an intensified use of tools. Representatives of the species *Homo* possessed everything that facilitated the spread into new habitats. Their physical characteristics allowed flexible locomotion for prolonged running as well as climbing. Due to the diversity of their diet, the requirements for a suitable habitat were lower. With their dexterity in handling various materials and tools, their social interactions, their increasing intellectual abilities, and, as a result of all this, their growing cultural possibilities, they were able to quickly adapt to new conditions. It was thus possible to advance into areas that did not exactly correspond to the African environment, and curiosity spurred them on to look over the next mountain.

4 Between 3 and 2 million years ago, two new genera evolved: *Paranthropus* and *Homo*. Some of them spread to northern Africa.

Fig. 4

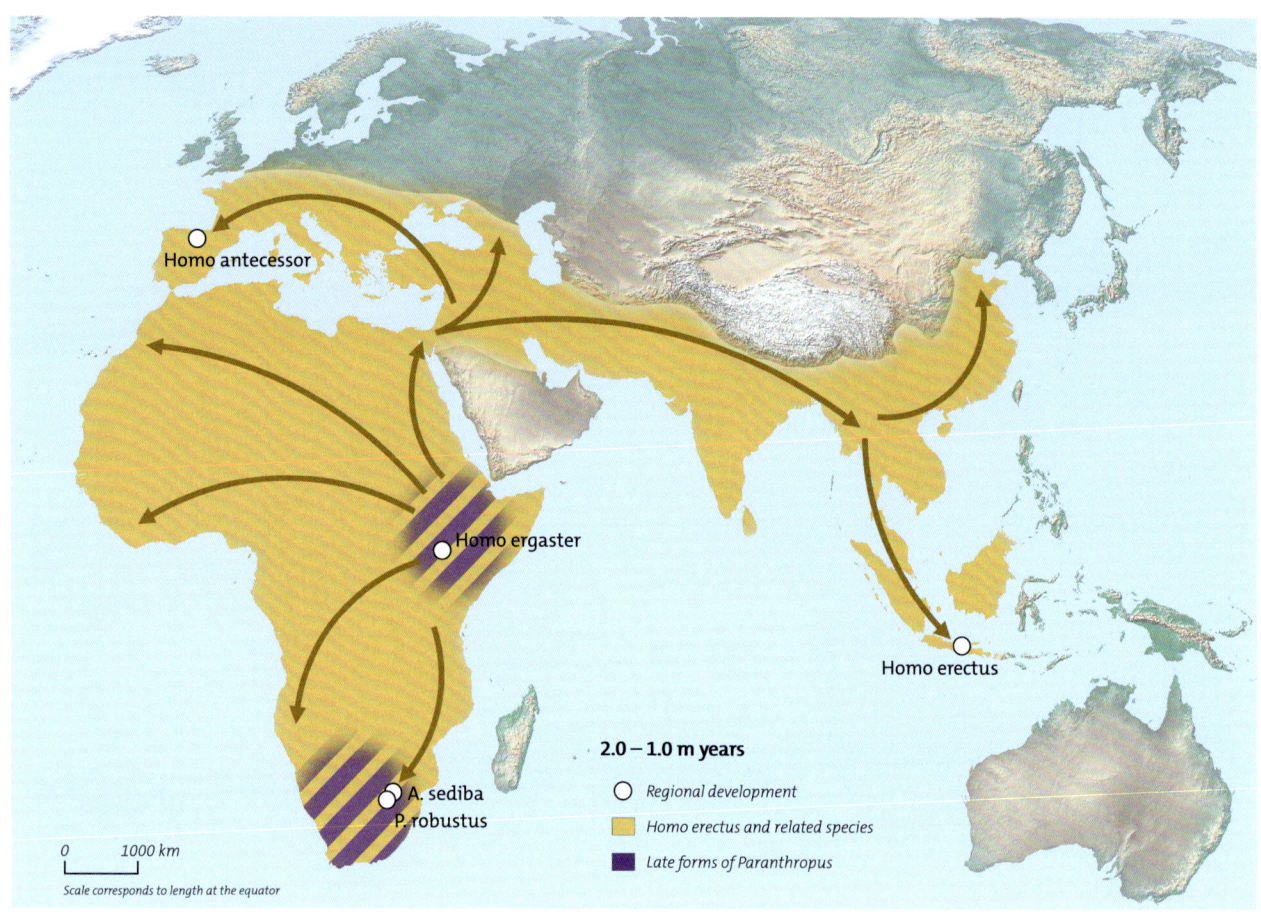

Fig. 5

5 Between 2 and 1 million years ago, *Homo erectus* and related species reached East Asia and Europe. *Australopithecus* and *Paranthropus* species still inhabited East and South Africa.

Recent archeological developments illustrate how dependent we are on individual finds and their dates when writing the history of human expansion. For almost 30 years, the 1.8 million-year-old Dmanisi site in Georgia with simple stone tools and human fossils was considered to represent the oldest evidence of the first spread of humans outside of Africa. Since 2018, the evidence for a much earlier expansion of *Homo* from their African core area is piling up: to North Africa with the up to 2.4-million-year-old Algerian site Ain Boucherit, to the Levant with the up to 2.48 million-year-old finds from the Jordanian Zarqa Valley, and—hardly later—to East Asia with the up to 2.1 million-year-old site of Shangchen in central China.

Due to the small number of finds and the approximate dating, it is difficult to say how fast humans spread into the areas outside of Africa that were previously uninhabited by humans. If we calculate an average shift in the explored territory of only 1 km per generation (approx. 20 years) in one direction, then it is easily possible to bridge 10,000 km in 200,000 years. Fossils and tool finds indicate a settlement of Java and northern China by *Homo erectus* from around

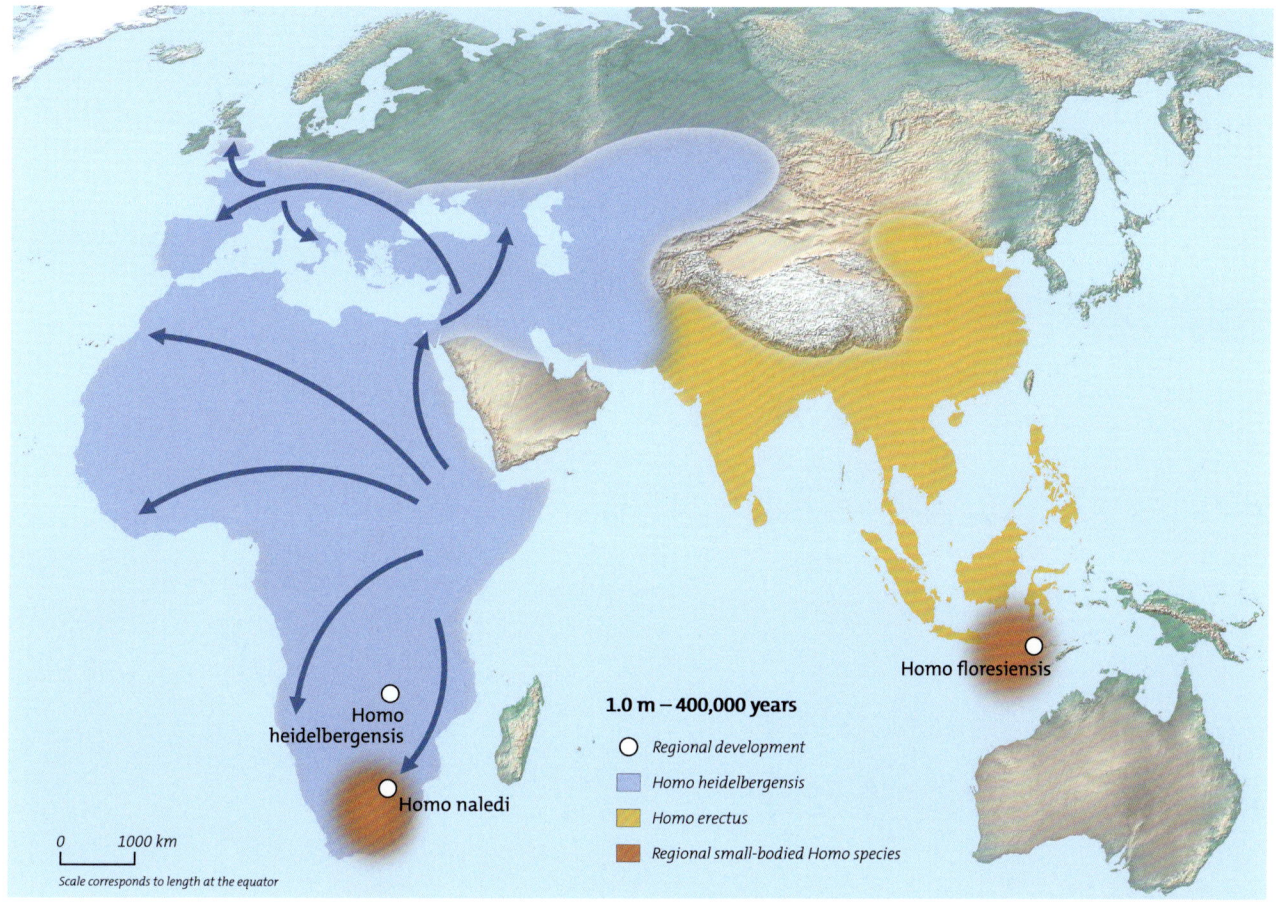

Homo floresiensis

Homo
heidelbergensis

Homo naledi

1.0 m – 400,000 years

○ *Regional development*

▢ *Homo heidelbergensis*

▢ *Homo erectus*

▢ *Regional small-bodied Homo species*

0   1000 km

Scale corresponds to length at the equator

1.6 million years ago. The first finds in Europe come from the Mediterranean area from a time up to 1.2 to 1.4 million years ago. It was probably not a single expansion, but a gradual process of several spreading and retreating episodes. For East Asia, as for Europe, it remains unclear whether the finds are evidence for permanent settlements or only indications of recurring advances in times of favorable environmental conditions.

6 Between one million and 400,000 years ago, *Homo heidelbergensis* spread from Africa to Europe and parts of Asia. Regional small-statured *Homo* species emerged in southern Africa and on the island of Flores in Southeast Asia.

Diverse migrations

The spread into previously uninhabited areas is relatively easy to grasp using both fossils and artifacts. But once humans reached a specific region, it becomes more difficult. Do the later remains belong to a newly immigrated group or descendants of the first settlers? Were the new tool forms developed over time or were they imported from another region of origin? The first finds of a new stone tool technology in East Africa are around 1.75 million years old. With bifacial

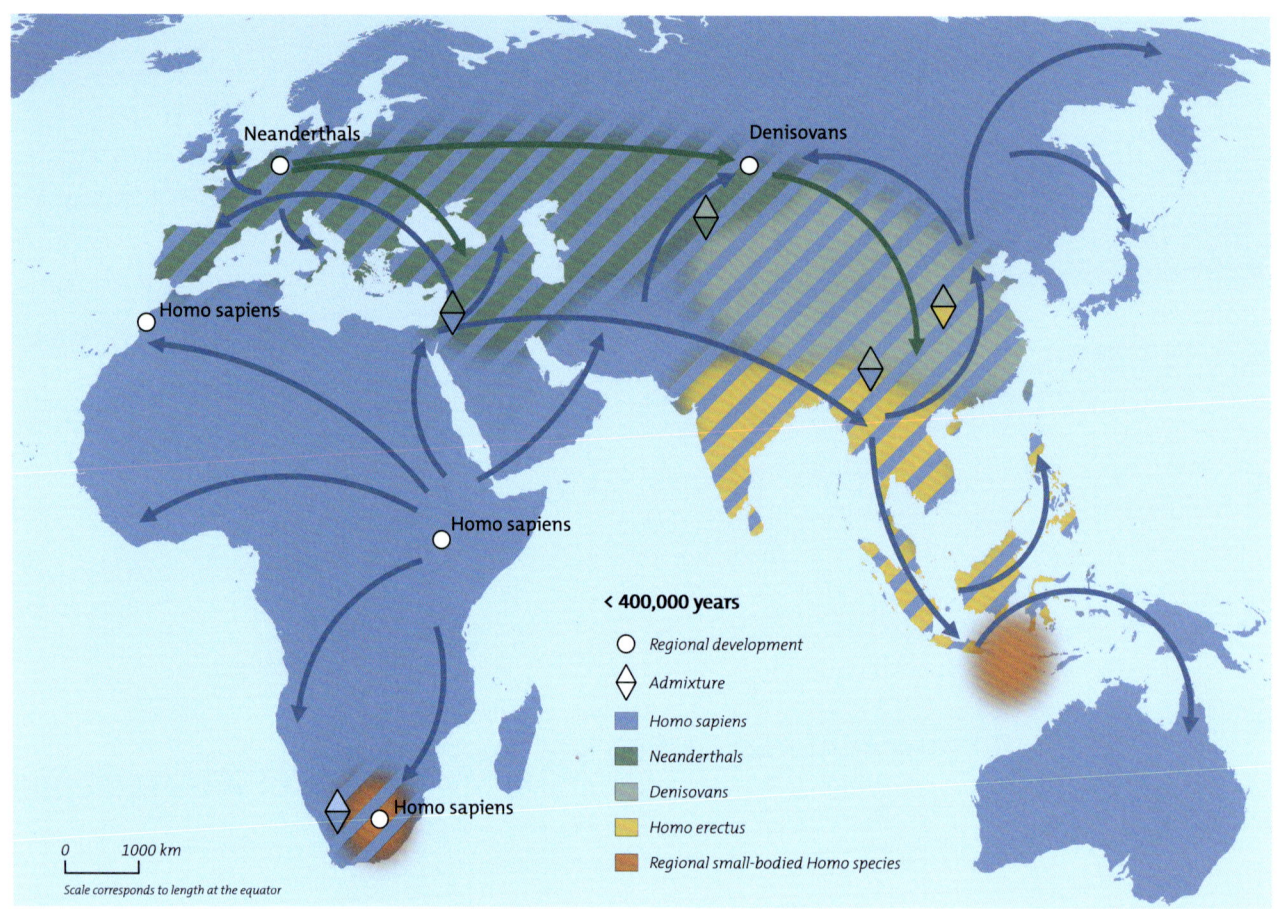

Neanderthals

Denisovans

Homo sapiens

Homo sapiens

< 400,000 years

○ Regional development

◇ Admixture

Homo sapiens

Neanderthals

Denisovans

Homo erectus

Regional small-bodied Homo species

Homo sapiens

0   1000 km

Scale corresponds to length at the equator

7 From 400,000 years ago Neanderthals developed in Europe, Denisovans in Asia, and *Homo sapiens* in Africa. Due to the different waves of expansions, it was not uncommon for them to mix. Today modern humans inhabit the whole earth.

Fig. 6

retouch of large blanks, *Homo* created easily manageable tools with coarse but stable edges: handaxes. Although this new, bifacial technique was used 1.5 million years ago at individual sites in the Levant and possibly in India, it only seems to have caught on around 600,000 years later in Asia and Europe. If this new technology was not invented independently in many places, the finds suggest a second and possibly also a third wave of expansion after one million years ago.

Global climatic fluctuations have intensified over the past one million years. Pronounced ice ages and warm periods alternated. The constantly changing conditions could hinder but also facilitate migrations. Deserts became impassable or greened, ice barriers arose and later gave way to dense forests, and vast cold steppes in Eurasia temporarily fed herds of large mammals. In many regions of the world, new human forms emerged that, like *Homo naledi* and *Homo floresiensis*, remained confined to small areas or spread, as is assumed for *Homo heidelbergensis* from Africa to Europe. Around 500,000 years ago, groups of people separated who later developed into Neanderthals in Europe and Denisovans somewhere in Asia. In Africa, *Homo sapiens* developed in parallel. New finds

118

from North Africa are more than 300,000 years old and include a mixture of old and new features, which suggest a slow development of our species, intertwined with the entire African continent.

Africa again and again

The history of the spread of *Homo sapiens* was probably not as short and linear as was long thought. Early genetic influences on European Neanderthal groups and fossil finds from Greece and Israel already indicate their presence outside Africa around 200,000 years before today. The further spread probably took place in several waves. Around 120,000 years before today there is evidence from the Arabian Peninsula, around 80,000 years before today they may have reached China via India, finds from Southeast Asia date to around 70,000 years before today. On their journey through a world inhabited by other species, *Homo sapiens* mixed with these groups again and again. From around 40,000 years ago, anatomically modern humans became the predominant human form in Eurasia. And more than 50,000 years ago they began to develop the remaining uninhabited areas for themselves: Australia, the Subarctic and Arctic, then North and South America, and finally the Pacific Island world.

Fig. 7

The complex history of humankind can only be understood from a global perspective. Each continent made own contributions and all tie into the overall development. Africa, however, stands out. The first human colonization of Eurasia came from Africa, all of today's humans have origins in Africa, and many impulses came from there. Why does Africa play this special role? Paul Bons and colleagues have an amazing mathematical answer to this. Without any special environmental or other external factors, their model demonstrates how large, relatively densely populated areas can develop into large waves of expansion, which are accompanied by numerous smaller waves of migration. For millions of years, Africa was the largest region with relatively dense human settlement, which statistically makes it the most likely place of origin of today's humans. It could have been that simple.

Further reading

Bellwood, P. (Hg.) 2013 The global prehistory of human migration (Chichester 2013).

Bons, P. D./Bauer, C. C./Bocherens, H./de Riese, T./Drucker, D. G./Francken, M./Menéndez, L./Uhl, A./van Milligen, B. P./Wißing, C. 2019 Out of Africa by spontaneous migration waves. PloS One 14(4), 2019, e0201998.

Groucutt, H. S./Petraglia, M. D./Bailey, G./Scerri, E. M. L./Parton, A./Clark-Balzan, L./ Jennings, R. P./Lewis, L./Blinkhorn, J./Drake, N. A./Breeze, P. S./Inglis, R. H./Devès, M. H./ Meredith-Williams, M./Boivin, N./Thomas, M. G./Scally, A. 2015 Rethinking the dispersal of *Homo sapiens* out of Africa. Evolutionary Anthropology 24, 2015, 149–164.

Homo sapiens neanderthalensis

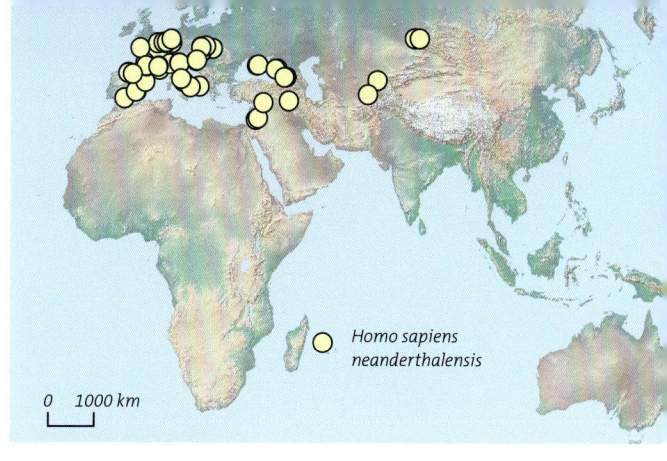

○ Homo sapiens
neanderthalensis

0   1000 km

Discovery

In 1829, Philippe-Charles Schmerling discovered fossils in a cave near Engis near Liège (Belgium). But it was not until 1856 that William King identified them as a unique human species based on the remains from the Feldhofer Grotto in the Neander Valley near Düsseldorf. This discovery, together with Charles Darwin's 1859 publication on the theory of evolution, called the Christian creation myth into question.

Sites

Europe: Belgium, Germany, France, Georgia, Italy, Croatia, Portugal, Spain, Czech Republic, Hungary, Poland, Romania.

Central Asia: Uzbekistan, Russia.

Middle East: Israel, Iraq, Syria, Turkey.

Finds

nearly complete skeletons, especially skulls, spinal column, shoulder blade, arm and leg bones.

Age

circa 175,000–30,000 years.

Brain size

circa 1,200–1,740 cm³.

Characteristics

Neanderthals are the best known fossil humans due to the recovered remains of more than 300 individuals. Anatomically, they differ little from humans today. Overall, they were of somewhat stronger and stockier. Their skulls were large and relatively long, they had heavy brow ridges over the eyes, a relative wide nose, and no chin. The shape of the hyoid bone is evidence that they had the anatomical prerequisites that would enable them to speak. Their diet was very variable: based on isotope analyses of teeth, researchers generally assumed a very high proportion of meat. Tests on dental tartar from Spanish fossils provided evidence of a rich vegetable diet. Neanderthals used a wide range of tools. They created wooden handles for stone tools using birch pitch, were probably able to make fire, and used red ochre dye. There is evidence that they buried their dead. Genetic studies show that Neanderthals mixed with both *Homo sapiens* and Denisovans.

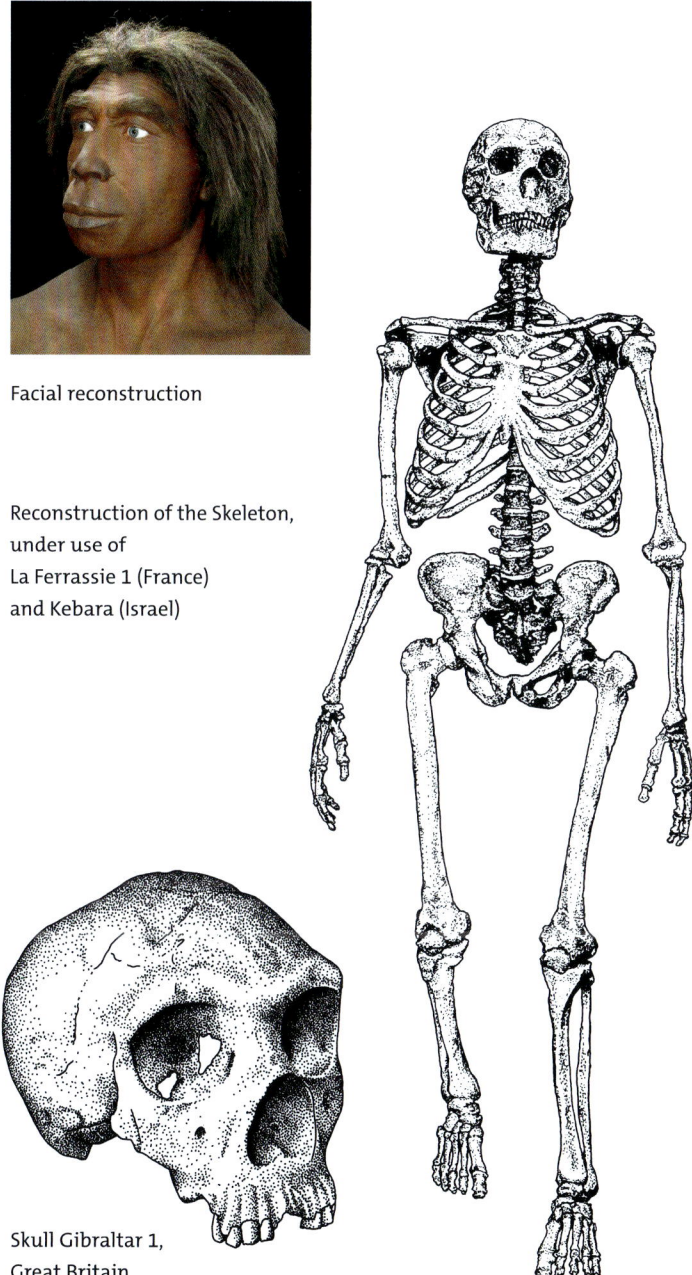

Facial reconstruction

Reconstruction of the Skeleton, under use of
La Ferrassie 1 (France) and Kebara (Israel)

Skull Gibraltar 1, Great Britain

Roman M. Wittig

Chimpanzee cultures — a search for clues

The search for chimpanzee cultures begins. If we start by asking www.wikipedia.de, our search comes to an abrupt end after the first sentence: "Culture, in the broadest sense, denotes everything that humans create or produce—in contrast to nature, which humans did not create and cannot alter." According to the authors of the article, culture is a human characteristic. The authors further argue that "the concept of culture can refer to a social group [...]. Commonalities of a group of people or the whole of humanity serve to distinguish this group from others or humans from animals." So, do animal cultures even exist? Can only humans be cultural beings?

In a biological sense, humans are animals. They are a species that has—undoubtedly—developed unparalleled skills. Despite this, they are simply a species of animal—*Homo sapiens*—whose own evolutionary line of development separated seven to eight million years ago from that of its closest living relatives, the chimpanzees *(Pan troglodytes)* and bonobos *(Pan paniscus)*, with whom it shares around 99 percent of the genetic make-up. In fact, chimpanzees are more closely related to humans than to gorillas. So, is a cultural adaptation to the environment something that only developed in the past five to six million years, after our evolutionary line separated from that of the chimpanzees?

In biology, we speak of cultural traditions when behavioral adaptations to the environment are not controlled by genetic (or epigenetic) processes but are passed on from one generation to the next through social learning. The sum of cultural traditions then becomes the culture of a group, which distinguishes it from others. Is this conceivable among our closest relatives, the chimpanzees? Now we begin our search in tropical Africa.

1 The mother cracks a nut with a heavy stone hammer while the offspring watches.

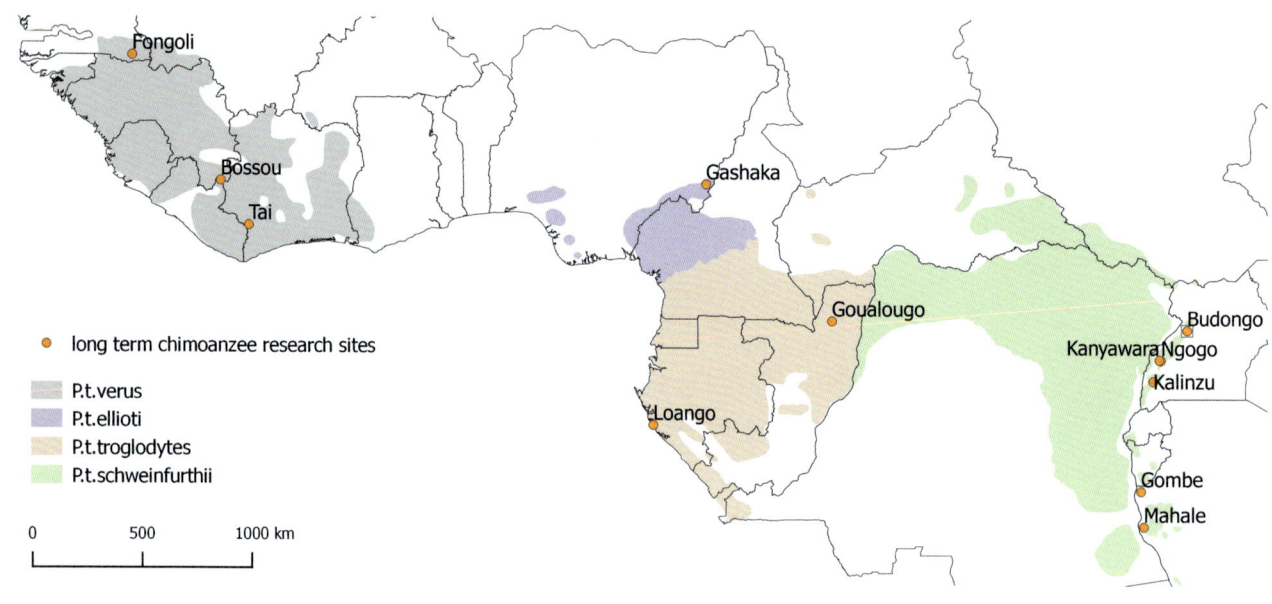

2 Distribution of the four subspecies of chimpanzees *(Pan troglodytes)* in Africa with the corresponding research projects that carry out or have carried out long-term research on chimpanzees.

Fig. 2

Chimpanzee ecology and tool use

The natural geographic distribution of chimpanzees extends over most of equatorial Africa, from Senegal in the west to Tanzania in the east. Four subspecies of chimpanzees are distinguished: *Pan troglodytes verus, Pan troglodytes elfioti, Pan troglodytes troglodytes*, and *Pan troglodytes schweinfurthii*. The habitat ranges from tropical rainforests to savannas with varying group sizes from ten to over 150 individuals. Chimpanzees are very territorial and live in mixed-sex fission-fusion communities. This means that the individuals in a group do not always remain together, but rather in changing subgroups, which form and rebuild at will—similar to human societies. Females give birth on average every five years. The juveniles normally stay with their mother until the onset of puberty, between the ages of ten and twelve, and thus develop at a similar rate as humans. When they reach sexual maturity (from around twelve years of age), the females join another group, while the males remain in their maternal group.

Chimpanzees are omnivorous. Although they mainly eat ripe fruits, they also consume meat, insects, nuts, mushrooms, leaves, honey, rotting wood, and much more. Most importantly, chimpanzees are extremely inventive when it comes to exploiting resources. To do this, they use a variety of tools, most of which the chimpanzees adapt to the task at hand:

(a) They use previously sharpened sticks as a spear to kill and eat nocturnal
galagos hiding in tree holes.

Fig. 3a–d

(b) They use sponges made from chewed-up leaves to soak up honey from a
hole in a tree trunk.

(c) They use stones or wooden clubs as hammers to crack open hard nutshells
according to the hammer-anvil principle to get to the inside of the nut
(see the Infobox Nutcracking). If part of the nut gets stuck in the shell, they
prepare a stick into the correct length to poke the remaining part of the
nut out of the shell.

(d) When fishing for termites, some chimpanzees use two different tools:
a sturdy stick to make a hole in the termite mound and a flexible twig to
then fish for termites through the hole.

Chimpanzees, therefore, use tools to exploit certain resources that they would
otherwise not reach or would have difficulty accessing. These tools must have
certain properties and are partially manufactured and modified. That's clever,
but can we call that culture? We should first consider whether there are differences between chimpanzee populations that are not based on genetics or
ecological factors.

Differences between populations

The variation of behavior was the first approach to take a closer look at the question of chimpanzee culture. This idea by Christophe Boesch and Andrew Whiten
was just as simple as it was ingenious: bring together researchers who have observed chimpanzee communities in the wild for many years and compile a precise
description of the observed behaviors. Then assign the observed behaviors to one
of three categories:

(I)   The first category includes universal behavior that all chimpanzees exhibited.

(II)   The second category includes behavior that not all chimpanzee groups
exhibited, which can be explained by ecological reasons. For example,
chimpanzees cannot crack nuts if there are no nuts available in their habitat.

(III) The third category includes behavior, the absence of which in chimpanzee
groups cannot be explained by ecological reasons, for example not cracking
nuts even though nuts and potential hammers and anvils are available.

Only category III behaviors are serious candidates for cultural traditions, while
category I and II behaviors are likely to be the result of genetic and ecological
adaptations. In the first study using such an approach in chimpanzees, the researchers identified 38 behaviors that could be classified as category III behavior

Fig. 4

(a) A sponge made from chewed-up leaves is used to extract water from a tree cavity.

(b) A wooden hammer is used to crack a coula nut using the hammer-anvil principle.

3 Four different tools in action.

by comparing six long-term field studies. These included behaviors such as cracking nuts, fishing for termites, or the rain dance. On the one hand, this study made it possible to investigate cultures among other great apes, on the other hand, a hefty dispute arose about whether ecological or genetic variations were, in fact, the real reason for the behavioral differences between the chimpanzee populations. So, are there really no behavioral differences based on cultural traditions?

Experimental approach

If simple observations do not provide answers, an experiment may help. But how to begin? An elegant approach was proposed by Thibaud Gruber from the working group around Klaus Zuberbühler. If behavioral differences are not based on cultural traditions, then a new problem presented to two different chimpanzee groups in an identical manner should provoke similar solution patterns. This is where the behavioral experiment began.

The researchers worked with two chimpanzee groups in Uganda. The Sonso group in Budongo Forest only uses leaves, not sticks, as tools. One could say that the Sonso chimpanzees live in a leaf culture. The Kanyawara group in Kibale

(c) A stone hammer is used to crack the hard shell of a panda nut.

(d) A toolset composed of a tough stick used to open the termite hill (right hand) and a flexible twig used to fish for the termites through the hole (in the mouth).

National Park about 200 km further south uses both leaves and sticks as tools. Researchers placed tree trunks with 16 cm deep holes filled with honey into the territories of both communities. To optimally exploit the honey, it should be spooned out of the hole using a sufficiently long stick. This is exactly what the Kanyawara chimpanzees did. The Sonso chimpanzees, on the other hand, who had never used sticks to extract food before, made a sponge out of leaves that they stuck into the hole and then pulled out again. Each group responded to the problem with their traditional techniques. Even when researchers demonstrated the optimal solution to the Sonso chimpanzees by placing a stick into the honey hole, the stick was cast aside, and leaves were used to extract the honey.

Fig. 5a

Fig. 5b

One possible explanation is that both techniques are equally suitable for obtaining honey and that there is no advantage of one technique over the other. Even if this is unlikely and there are clear advantages of using one technology instead of another in terms of how long the chimpanzees had to work to exploit the honey, there are still doubts. So, given the same effectiveness, do chimpanzees simply stick to the behavior variation that is more familiar to them?

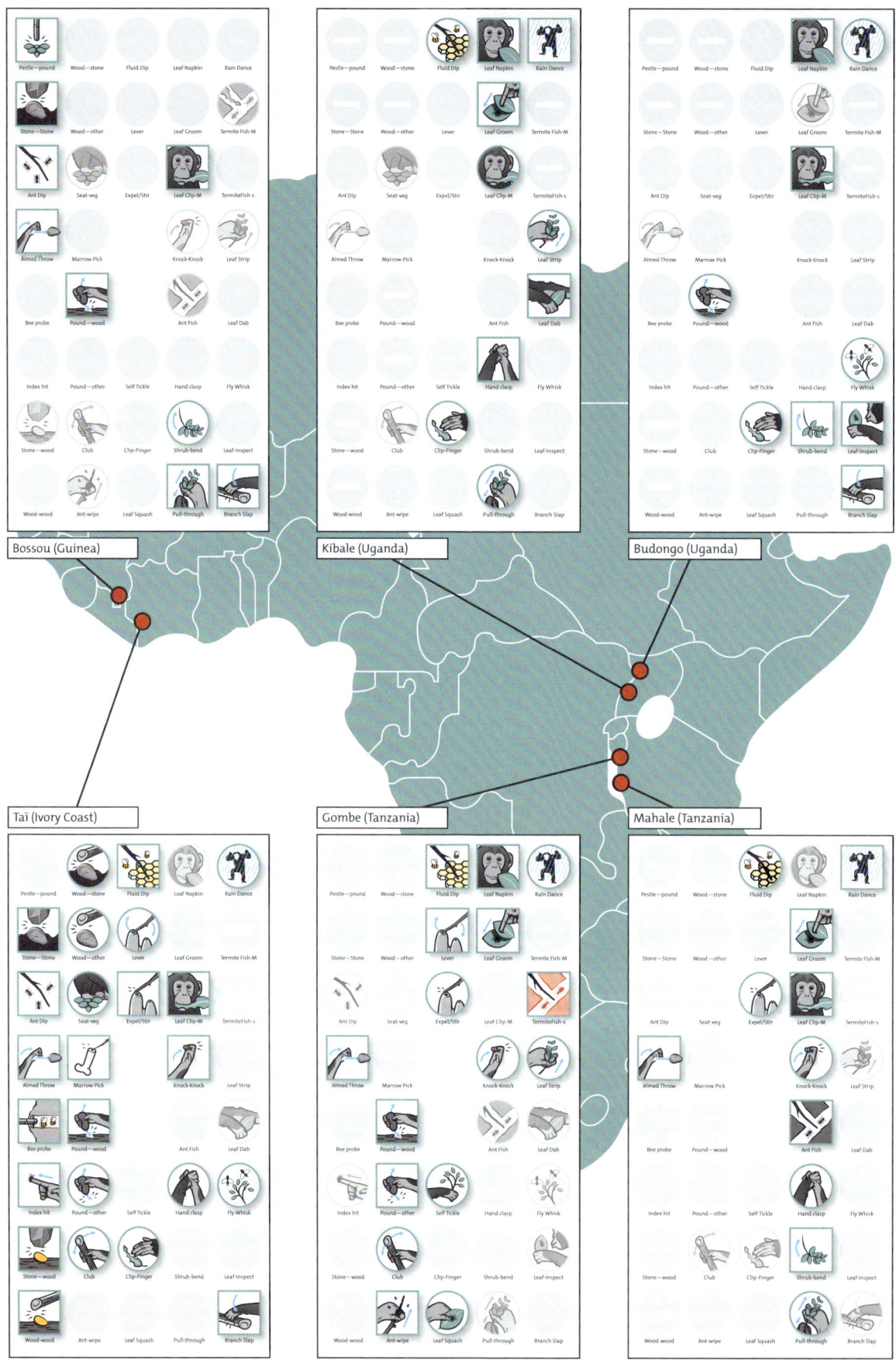

Bossou (Guinea)

Kibale (Uganda)

Budongo (Uganda)

Taï (Ivory Coast)

Gombe (Tanzania)

Mahale (Tanzania)

Adaptation to the cultural tradition of a new community

Here again, observations help us. Chimpanzees in the Taï National Park in the Republic of Côte d'Ivoire crack coula nuts *(Coula edulis)* from December to March. In December, when the fruits are fresh and hang on the trees, the chimpanzees use stone hammers to crack the hard nutshells. Later in the season once the fruits fall from the trees and begin to dry out, it is easier to crack the shells, so the chimpanzees now use wooden hammers. This makes sense since stone hammers are rare, while wooden hammers are readily available in the forest. In a typical chimpanzee community in Taï, the proportion of stone hammers changes over the course of the nutcracking season from 60 to 70 percent in the early weeks to 10 to 20 percent at the end of the nutcracking season, i.e., after around 20 weeks.

But there are also Taï chimpanzees who continue to prefer stone hammers. Lydia Luncz from Christophe Boesch's working group studied three neighboring communities, two of which follow the ecologically prescribed switch from stone to wooden hammers, while the southern group continues to use stone hammers in the twentieth week. They do this, although stones are not more numerous in the south than in the territory of the northern or eastern group. It looks as though two different hammer selection cultures exist: an alternating hammer selection culture, depending on how easy the nuts are to crack, and a stone hammer culture, which always uses stone hammers regardless of the degree of hardness of the nuts.

But what happens if a female from a hammer selection culture migrates to join the southern group with the stone hammer culture? Such migration was observed by our team in Taï National Park a few years ago. Within a few nutcracking seasons, the immigrant females had adapted to the predominant local culture and cracked the nuts according to the stone hammer culture predominant among the southern group. This cultural conformism in wild chimpanzees, which has already been shown in behavioral experiments with zoo chimpanzees, clearly shows that female chimpanzees adapt to their new culture.

Fig. 6

4 Distribution of the 38 behaviors observed during the six long-term research projects identified as possible cultural traditions by the authors.

Rectangular symbol:	common behavior (all or most of the members of a group display this behavior)
Round symbol:	frequent behavior (exhibited by several members of the group)
Pale gray symbol:	behavior present (but rarely observed)
Symbol without a picture:	behavior was not observed
Symbol without a picture with a bar:	behavior does not exist for ecological reasons

Conclusion

Our search for chimpanzee culture has shown that the foundations of human culture do already exist among them, even if chimpanzees are not building cities, singing operas, or flying to the moon. Chimpanzees shape sticks and leaves into functional tools and intentionally use stones as hammers. The use of tools is learned and sometimes takes many years to perfect. Some of these behaviors occur in one community, but not in another, for no apparent genetic or ecological reason. Ultimately, chimpanzees adapt to a new predominant culture, even if it would be more effective to persist in their old culture.

Perhaps the result of our search shows that we must accept that the chimpanzees do have their own culture. This only seems logical when we see how long it takes for chimpanzee children to reach certain developmental stages and that the use of tools is not easily learned on an individual basis. For some feats, chimpanzees need a role model from whom they can learn—such as the mother with whom the juveniles roam the jungle for years.

Acknowledgments

I would like to thank the Ministère de l'Enseignement Supérieur et de la Recherche Scientifique, the Ministère des Eaux et Forêts in Côte d'Ivoire, and the Office Ivoirien des Parcs et Réserves for forty years of research into chimpanzee behavior in the Taï National Park. Thanks also to the Centre Suisse de Recherches Scientifiques en Côte d'Ivoire and to the staff of the Taï Chimpanzee Project for their ongoing collaboration. My special thanks go to Christophe Boesch, Cedric Girard-Buttoz, Thibaud Gruber, Lydia Luncz, Dave Morgan, Crickette Sanz, and Liran Samuni for discussions about tool use and providing illustrations.

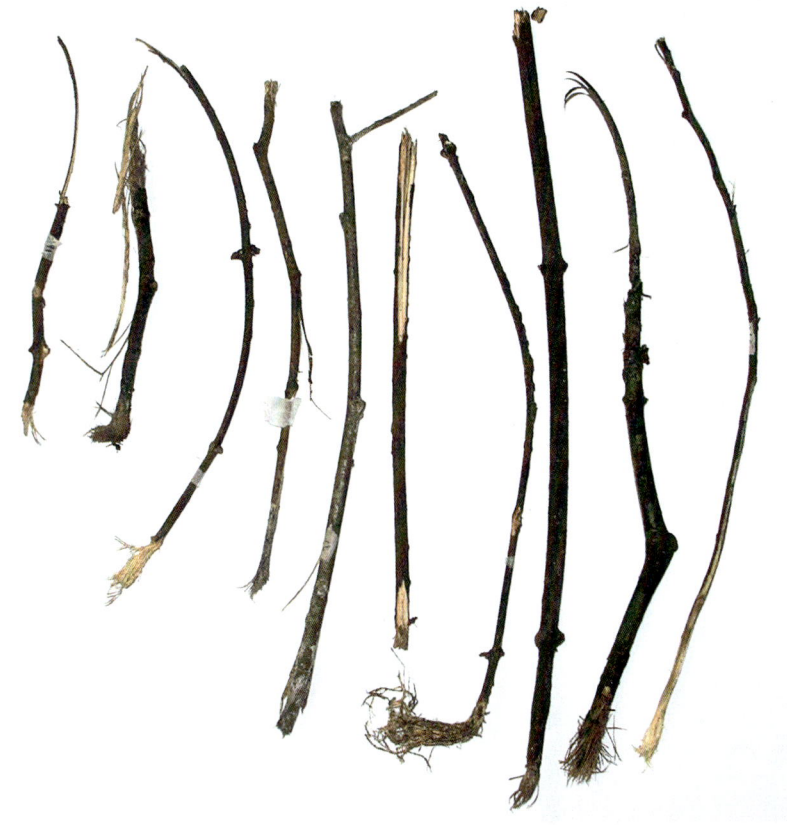

5 (a) Sticks used by the Kanyawara chimpanzees to extract honey from a tree cavity.

(b) Leaf sponges used by the Sonso chimpanzees to solve the same problem: extracting honey from a tree cavity.

Further reading

Boesch, C. 2012 Wild cultures – a comparison between chimpanzee and human cultures (Cambridge 2012).

Boesch, C./Wittig, R.M. (Hg.) 2019 The chimpanzees of the Taï forest: 40 years of research (Cambridge 2019).

Estienne, V./Cohen, H./Wittig, R.M./Boesch, C. 2019 Maternal influence on the development of nut-cracking skills in the chimpanzees of the Taï forest, Côte d'Ivoire (Pan troglodytes verus). American Journal of Primatology 81(7), 2019, e23022. https://doi.org/10.1002/ajp.23022

Estienne, V./Stephens, C./Boesch, C. 2017 Extraction of honey from underground bee nests by central African chimpanzees (*Pan troglodytes troglodytes*) in Loango National Park, Gabon: Techniques and individual differences. American Journal of Primatology 79(8), 2017, e22672. https://doi.org/10.1002/ajp.22672

Gruber, T./Muller, M.N./Strimling, P./Wrangham, R./Zuberbühler, K. 2009 Wild chimpanzees rely on cultural knowledge to solve an experimental honey acquisition task. Current Biology 19 (21), 1806–1810. https://doi.org/10.1016/j.cub.2009.08.060

Humle, T./Matsuzawa, T. 2002 Ant-dipping among the chimpanzees of Bossou, Guinea, and some comparisons with other sites. American Journal of Primatollogy 58(3), 2002, 133–148. https://doi.org/10.1002/ajp.10055

Luncz, L.V./Mundry, R./Boesch, C. 2012 Evidence for cultural differences between neighboring chimpanzee communities. Current Biology 22(10), 2012, 922–926. https://doi.org/10.1016/j.cub.2012.03.031

Luncz, L.V./Wittig, R.M./Boesch, C. 2015 Primate archaeology reveals cultural transmission in wild chimpanzees (*Pan troglodytes verus*). Philosophical Transactions of the Royal Society B: Biological Sciences 370, 2015, 20140348. https://doi.org/10.1098/rstb.2014.0348

Pruetz, J.D./Bertolani, P. 2007 Savanna chimpanzees, *Pan troglodytes verus*, hunt with tools. Current Biology 17(5), 2007, 412–417. https://doi.org/10.1016/j.cub.2006.12.042

Nutcracking

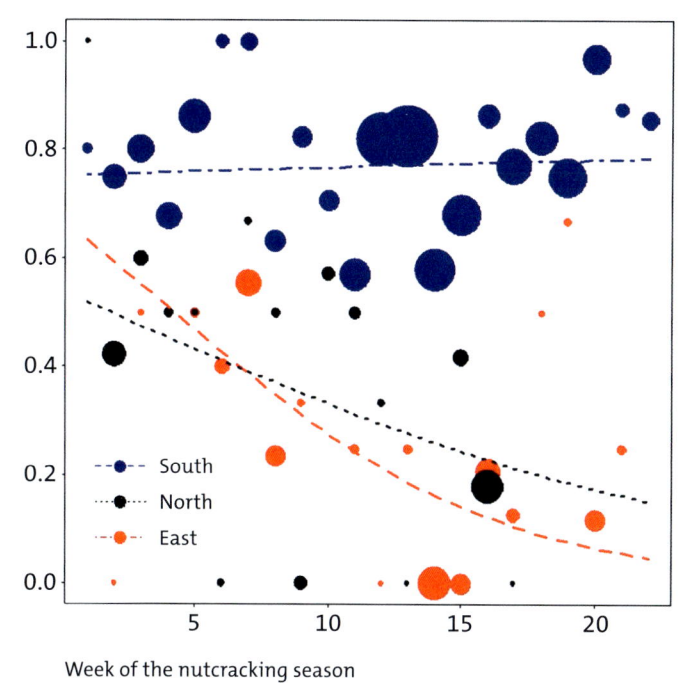

"Bum, bum—crack" echoes through the Taï National Park along the Ivory Coast on the border with Liberia. Two thumps and a loud crack. A group of chimpanzees is sitting under a Panda oleosa tree, cracking the hard-shelled seeds of the panda nuts to reach the soft inside core. Again, "bum, bum—crack". This time, a mature female chimpanzee cracked open one of the hard nuts. The female carefully lifts a large stone, weighing around 7 to 8 kg, about 30 to 40 cm with both hands and a foot, and then drops it like a hammer to crack the nut with one forceful blow (see Fig. 3c). A dull "bum" resounds. But the nut is still not open; a heavy hammer is not enough; she needs a corresponding anvil. The combination of the force of the blow and the matching strength of the anvil is required to crack the nut. She takes the nut from the anvil and, with the back of her hand, pushes aside the remains of the soft orange pulp of the panda fruit and places the nut back into the small hollow that the hard shells of previous panda nuts have carved into the root after years of use as an anvil. She lifts the stone and forcefully drops onto the nut—"bum"—and again—"bum"—and again—"crack"! Finally, the shell bursts. She puts the stone aside and brings part of the opened nutshell to her mouth to pull the soft white nutmeat out of the shell with her finger and tongue. At the age of five, her daughter is still too young to handle the heavy hammer herself. She lacks the necessary precision and strength. She sits across from her mother and watches every movement with great interest. Cracking nuts is not an easy matter. It takes many years of learning, and some Taï National Park chimpanzees do not become efficient nutcrackers until they are ten.

6 Proportion of stone hammers (vertical: the number of stone hammers divided by the number of all hammers) used by the Taï chimpanzees to crack open coula nuts throughout the nutcracking season (weeks). The groups from the north and east use fewer stone hammers over the course of the season as the nutshells become softer, while the group from the south continues to use stone tools (80 percent) throughout the season.

View the videos on Taï Chimpanzee Project Youtube-channel: https://www.youtube.com/channel/UC1tvBgBAV5Xlmm5Vh5GwUlw/videos.

In search of the beginnings of culture
at Lake Manyara in Tanzania.

Homo sapiens

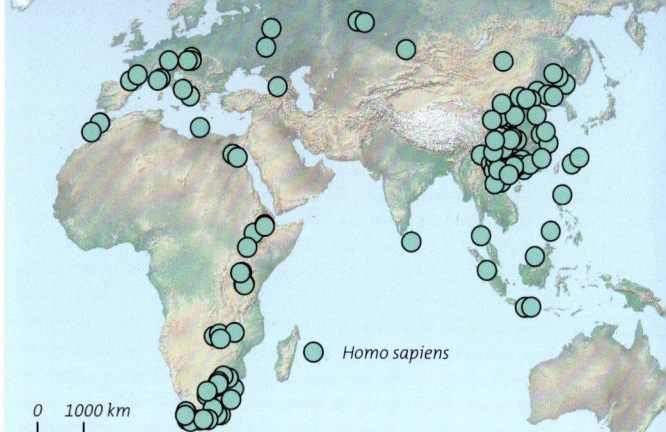

Homo sapiens

0 1000 km

Discovery

In 2017, the circa 300,000 year old, and therefore also the oldest-known, fossil remains of an anatomically modern human were discovered at Jebel Irhoud in Morocco. Until this discovery, the 195,000 year old skeletal finds from the Omo Valley in Ethiopia were the oldest-known representatives of *Homo sapiens*.

Spread

Worldwide

Age

since circa 300,000 years.

Brain size

circa 1,100–1,900 cm³ (on average circa 1,350 cm³).

Characteristics

Although *Homo sapiens* is the only species of all hominins still living today, genetic studies show small proportions of genes from Neanderthals, Denisovans, and other ancient humans in our genome. Our skulls are characterised by the relatively small, vertical face, a high forehead, a protruding chin, and a large cerebral skull. The skeleton has long leg bones, an opposable thumb for fine motor tasks, and a barrel-shaped chest. The s-shaped spine and the slightly tilted pelvis perfect the bipedal locomotion. Anatomically modern humans were the first to colonise the whole world, including Australia, the Arctic, North and South America, and Oceania. The extremely variable use of tools allows humans living today to enjoy very different forms of lifestyle, nutrition, and resource use. As a species, we are capable of unprecedented environmental changes, but also of cross-regional to global cooperation.

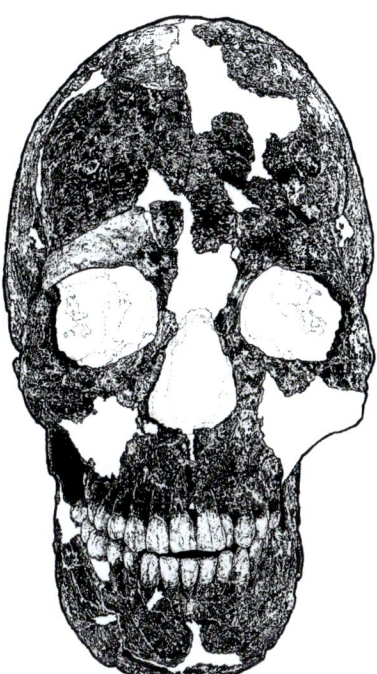

Skull Qafzeh IX, Israel

Virtual skull reconstruction
of Jebel Irhoud, Morocco

Madelaine Böhme

Tracing the steps to becoming human: research and scientific methods

The process of becoming human after the divergence of the human evolutionary line from that of the chimpanzees goes back many millions of years. It begins long before the first known cultural products such as tools. Over the past decades, a variety of methods were applied or specially developed to decipher this process. A complex system of analytical methods has emerged with broad involvement of the natural sciences, including geology, biology, chemistry, physics, and engineering. We present some of these fascinating ways to gain insight into our past below.

The **time dimension** is the most important context of historical research. The interpretation of data and observations is impossible without a temporal classification, as this is the only way to identify sequences and separate causes from effects. A distinction must be made between relative and absolute dating techniques using stratigraphy (geological methods that describe the sequence of layers) and geochronology (physical methods). Relative dating determines the sequence of two events (or the time two objects were created) and the relative timespan between them. Absolute dating indicates a date for an object. In the time depths of interest here, such a date is determined using various methods, such as the radioactive decay of certain elements, and expressed with age ranges. Sometimes it is more important to know whether a fire occurred right before a new settlement was built rather than to know that both events occurred 6,832 +/- 65 years ago, without a clear order of events.

Since becoming human is also a biological process, the scientific discipline of biology, including anatomy, physiology, genetics, zoology, and botany, plays a significant part in this research. The **evolutionary relationship** of our ancestors to one another and specifically *Homo sapiens* can be examined through comparative morphology (analysis of shapes and patterns) of the preserved bones and

1 Sampling of a rib for radiocarbon dating
and determination of the carbon-nitrogen ratio.

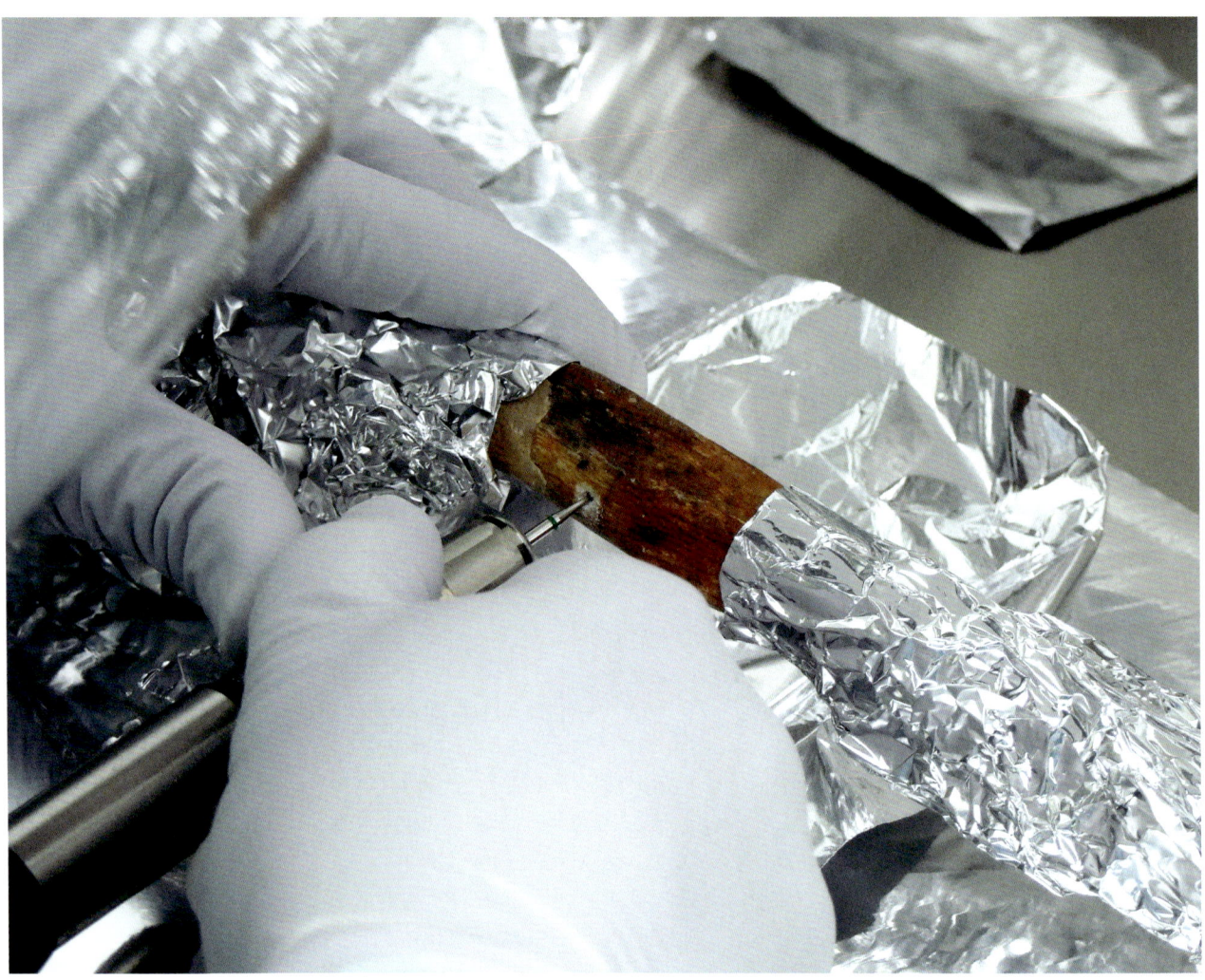

2 Taking a bone sample
for genetic analysis.

Fig. 2

teeth. Pronounced brow ridges above the eyes characterize the Neanderthals, but—expressed in a slightly different form—also other early human species, whereas they no longer occur in modern humans. If bones and teeth still contain organic substances such as collagen, an analysis of the preserved genetic fragments or proteins can sometimes provide very detailed information on evolutionary and individual relationships. The Denisova humans were recognized as a distinct species based on genetic snippets extracted from a fingertip bone. And the genetic material extracted from the bone fragment of a girl who lived in Siberia around 90,000 years ago provided evidence that her mother was a Neanderthal woman and her father a Denisovan!

Comparative morphology provides numerous clues about the **life history** of individuals and populations. By examining the different age stages of a specific human type, it is possible to obtain information about growth and development, and the length of childhood. We now know that Neanderthals and all our ancestors had much shorter childhoods and became adults much earlier. We can learn a lot about their developmental biology and sociology from this. By estimating age at the time of death, we can also roughly determine the mean life expectancy.

We can examine important aspects of **reproductive biology** using comparative morphology. The anatomy of our female ancestors' pelvis is different from ours. Australopithecines did not have a narrowed birth canal, so the babies did not have to rotate during birth. The babies came into the world much more mature, that is, less helpless and dependent on their mother, since their smaller heads did not require early births. It is possible to determine how long a child was breastfed based on the ratio of the elements calcium and strontium in tooth enamel. During the formation of tooth enamel, these elements are incorporated into the tooth depending on their prevalence in the food. Mother's milk is made up of a different ratio of these elements than subsequently consumed food. It was found that Neanderthal children were given solid foods from the age of 5–6 months.

Fig. 3

References to the **social structure** and **social behavior** of our ancestors could also be encoded in their bones. Hormonal predispositions change bone growth and influence the social behavior of a species. If male and female individuals differ only slightly in anatomy and body size, and the length of the index and ring fingers of their hands are the same, this indicates monogamous relationships. Modern gibbons are an example of this. If, on the other hand, male individuals are significantly larger and stronger than females in terms of body mass, canine teeth, and muscle attachments, and the index fingers are significantly shorter than the ring fingers in both sexes, this indicates polygyny, that is, harem behavior, as we observe among gorillas and chimpanzees. Our modern human behavior lies between the two extremes, albeit much closer to the gibbons.

Who has not heard the saying: "Do I have to chew your food for you too?"! The evidence of toothless old individuals who could no longer chew their food without help from others (for example by pre-chewing) provides important information about social bonds and altruistic behavior. This truly human behavior, unknown in the animal kingdom, was first documented for *Homo erectus,* 1.8 million years ago. We can also draw conclusions about social behavior from geoarchaeological studies of dwellings and social spaces such as hearths and sleeping quarters.

Bone modifications indicate injuries, illnesses, or malnutrition, which in turn have a lot to do with **living conditions** and **diets**. The diet is an important mirror of our living conditions. Analyses of the anatomy of teeth and jaws, as well as their diseases, not only provide us with information on behavior (for example using teeth as tools, using toothpicks, smoking pipes) but especially on the type of food consumed. We can interpret whether it was tough or firm, like dried

Fig. 1

meat or vegetable roots, or soft, like porridge or fast food, from the wear on the teeth, from their position and structure. Very sugary food can lead to tooth decay, and the different nutritional properties provided by plants, fish, and meat are stored as different isotope ratios in the bones. Food remains are preserved in rare instances, for example as microscopic traces on tools or vessels. In such cases, chemical analyzes can help identify the use of blood or milk or even the production of wine or cheese.

That brings us to the **preparation of the meal**. The control and use of fire and thus the ability to cook, bake, or grill is of crucial importance for human evolution. In many cases, it only made it possible to digest certain foods and absorb their nutrient in the first place, for example by detoxifying and changing the consistency, or at least decisively improving the experiences, as in the case of starchy plants. Due to the thermal pretreatment of food, the (cooking) human is the only mammal with the ability to (pre-)digest, outside of the stomach. Evidence for the use of fire was discovered by analyzing charcoal and hearths and also through observing chemical or physical changes of heated stones or floor surfaces.

The procurement of food is dependent on **hunting and gathering**, and later **farming and agriculture**. Cut and impact marks on bones are evidence of the use of meat or bone marrow. Hunting tools are rarely preserved as well as the oldest-known weapons in human history, the approximately 300,000-year-old wooden spears and throwing sticks associated with *Homo heidelbergensis* from Schöningen. Plant residues rarely survive in the archeological record, which is why less can be said about the plant-based diet in the Paleolithic. Analyses of the dental tartar of Neanderthal teeth identified plant remains from a wide range of foods.

Technological development is another, extensive research field of human evolution. It is not limited to material aspects of organic raw materials such as wood and bones or inorganic materials such as stone and pigments, but also includes the spatial distribution of objects and constructions, as they arise from the processing of materials or the organization of settlements.

Raw materials, in turn, provide us with evidence for **economic and ritual action**. The origin of organic and inorganic raw materials reveals a lot about supra-regional relationships and knowledge transfer. By analyzing devices, art objects, and burials, we learn something about the values and beliefs of early communities.

3 Comparative morphology helps reconstruct the life story of a person. Modern scanning techniques facilitate this process.

Further reading

Böhme, M./Braun, R./Breier, F. 2019 Wie wir Menschen wurden: Eine kriminalistische Spurensuche nach den Ursprüngen der Menschheit (München 2019).

Hauptmann, A./Pingel, V. 2008 Archäometrie (Stuttgart 2008).

Krause, J./Trappe, T. 2019 Die Reise unserer Gene: eine Geschichte über uns und unsere Vorfahren (Berlin 2019).

Meller, H./Alt, K. W. (Hg.) 2010 Anthropologie, Isotopie und DNA – biografische Annäherung an namenlose vorgeschichtliche Skelette? 2. Mitteldeutscher Archäologentag vom 08. bis 10. Oktober 2009 in Halle (Saale). Archäologie Sachsen-Anhalt 3 (Halle/Saale 2010).

Would you like to learn more about individual sites?

The research center ROCEEH (The Role of Culture in Early Expansions of Humans) is a project of the Heidelberg Academy of Sciences and Humanities with the aim of exploring the early cultural heritage of humankind, placing it in context, and preserving it. ROCEEH explores the history of humankind and its early spread from three million to 20,000 years ago. By compiling archeological sites and the information associated with them, ROCEEH makes the earliest cultural heritage accessible.

Systematically collected data from sites in Africa and Eurasia are archived in the ROCEEH Out of Africa Database (ROAD). This contains a variety of archeological, paleoanthropological, paleobiological, geographic, and bibliographical information. As of early 2022, ROAD contains data from around 2,200 sites and 17,000 inventories. Information on each of these sites can be accessed as ROAD Summary Data Sheets as a PDF without registration at the following address: https://www.roceeh.uni-tuebingen.de/roadweb/ [19/03/2022]. If you would like to learn more about the ROCEEH research center and other analysis options involving the ROAD database, please visit the project homepage: www.roceeh.net [19/03/2022].

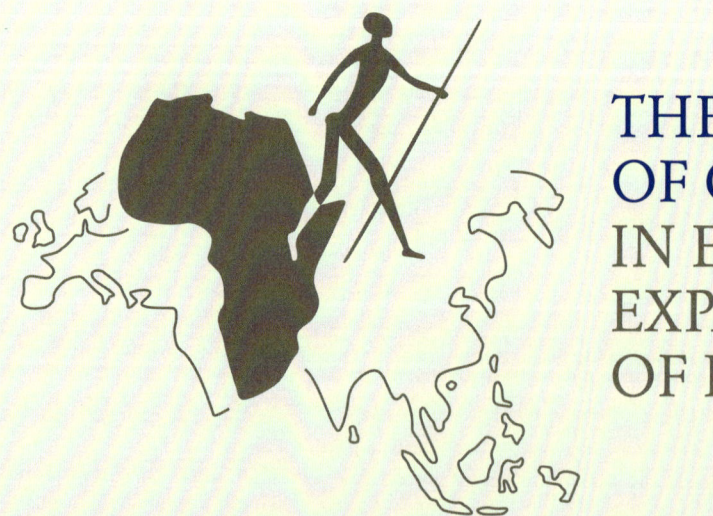

THE ROLE OF CULTURE IN EARLY EXPANSIONS OF HUMANS

Author Adresses

Prof. Dr. Madelaine Böhme
Eberhard Karls University Tübingen
Senckenberg Center for Human Evolution
and Palaeoenvironment (HEP Tuebingen)
Sigwartstr. 10
72076 Tübingen, Germany
m.boehme@ifg.uni-tuebingen.de

apl. Prof. Dr. Michael Bolus
Research Center 'The Role of Culture in
Early Expansions of Humans' (ROCEEH)
at Tübingen University, Germany
Hölderlinstraße 12
72074 Tübingen, Germany
michael.bolus@uni-tuebingen.de

PD Dr. Angela A. Bruch
Research Center 'The Role of Culture in
Early Expansions of Humans' (ROCEEH)
at Senckenberg Research Institute
and Nature Museum
Senckenberganlage 25
60325 Frankfurt am Main, Germany
angela.bruch@senckenberg.de

Dr. Liane Giemsch
Archäologisches Museum Frankfurt
Karmelitergasse 1
60311 Frankfurt am Main, Germany
liane.giemsch@stadt-frankfurt.de

Dr. Karen Hahn
Institute for Ecology, Evolution, and Diversity
Goethe University Frankfurt
Dept. Paleobiology and Environment
Max-von-Laue-Str. 13
60438 Frankfurt am Main, Germany
karen.hahn@bio.uni-frankfurt.de

PD Dr. Miriam Noël Haidle
Research Center 'The Role of Culture in
Early Expansions of Humans' (ROCEEH)
at Senckenberg Research Institute
and Nature Museum
Senckenberganlage 25
60325 Frankfurt am Main, Germany
miriam.haidle@senckenberg.de

Prof. Dr. Thomas Junker
Skylineblick 14
60438 Frankfurt am Main, Germany
mail@tjunker.de

Dr. Christine Michel
Pedagogical Faculty
University Leipzig
Early Child Development and Culture
Jahnallee 59
04109 Leipzig, Germany
christine.michel@uni-leipzig.de

PD Dr. Oliver Schlaudt
Philosophical Seminar of University Heidelberg
Schulgasse 6
69117 Heidelberg, Germany
oliver.schlaudt@urz.uni-heidelberg.de

Prof. Dr. Friedemann Schrenk
Institute for Ecology, Evolution, and Diversity
Goethe University Frankfurt & ROCEEH, Senckenberg
Research Institute and Nature Museum
Senckenberganlage 25
60325 Frankfurt am Main, Germany
schrenk@senckenberg.de

Dr. Roman M. Wittig
Taï Chimpanzee Project
Max Planck Institute for Evolutionary Anthropology
Deutscher Platz 6
04103 Leipzig, Germany
wittig@eva.mpg.de

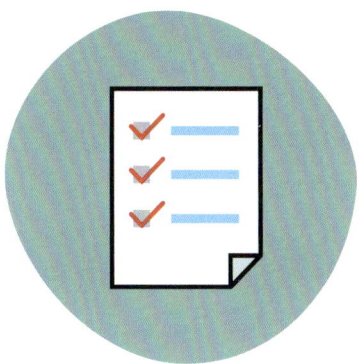

Picture credits

AMF = Archäologisches Museum Frankfurt

Cover: Design: E. Quednau/AMF;
Illustration: B. Groscurth/AMF/ROCEEH
Front inside cover:
Design: B. Groscurth/AMF/ROCEEH
Busts: Atelier Wildlife Art for Hessisches
Landesmuseum Darmstadt, Photos:
W. Fuhrmannek, HLMD, Drawing: *Ar. Ramidus,
P. aethiopicus, H. floresiensis, H. antecessor,
H. heidelbergensis, H. sapiens:* A. Marie Rahn;
ER 3733: Christine Hemm
back: L. Giemsch

Initial icons for contributions:
B. Groscurth/AMF/ROCEEH

Liane Giemsch, Miriam Noël Haidle
In search of the beginnings of our culture
1 Atelier Wildlife Art for Hessisches
Landesmuseum Darmstadt,
Photos: W. Fuhrmannek, HLMD.
2 B. Groscurth/AMF/ROCEEH

Thomas Junker
**Between nature and culture:
the two origins of humanity**
1 Tyson, E. 1699 Orang-Outang, sive *Homo
Sylvestris:* or, the Anatomy of a Pygmie
compared with that of a Monkey, an Ape,
and a Man (London 1699), Abb. 1.
2 Belon, P. 1555 L'histoire de la nature des
oyseaux (Paris 1555), 40–41.
3 Junker, T. 2018 Die Evolution des Menschen.
3. Aufl. (München 2018), 22.
4 Haeckel, E. 1911 Natürliche Schöpfungs-
geschichte. 11. Aufl. (Berlin 1911), Tafel 29.
5 Bonner, J. T. 1980 The evolution of culture in animals (Princeton 1980) 1980, 167.

Friedemann Schrenk
Early human biocultural evolution
1 Friedemann Schrenk
2 A. M. Rahn, Ch. Hemm
3 Ch. Hemm
4 A. M. Rahn; American Museum of
Natural History
5 Black skull: A. M. Rahn
KNM-ER 1470: Ch. Hemm
6 Drawings: A. M. Rahn; Virtual skull
reconstruction *Homo sapiens*, Max Planck
Institute for Evolutionary Anthropology,
Leipzig; Zhkoudian: skull reconstruction,
American Museum of Natural History
7 Hominin reconstructions: WildLifeArt
N. Kieser & W. Schnaubelt, Photo: F. Schrenk

Michael Bolus
**The earliest stage of human stone tool
technology: the Oldowan**
1 Didier Descouens. Reproduction under
the Creative Commons Licence CC BY-SA 4.0
(https://commons.wikimedia.org/
wiki/File:Pierre_taill%C3%A9e_Melka
Kunture%C3%89thiopie_fond.jpg#/
media/File:Pierre_taill%C3%A9e_Melka
Kunture%C3%89thiopie.jpg). Accessed
14.09.2020.
2 B. Groscurth/AMF/ROCEEH
3 Taken from Haidle 2012 (there compiled
from Leakey 1971).
4 Data: ROAD and Michael Bolus
(ROCEEH), Background map: T. Patterson,
used under 'public domain' licensing,
cartography: Christian Sommer (ROCEEH).

5 According to Harmand 2007 (see also
Delagnes/Roche 2005).
6 Data: ROAD (ROCEEH), background map:
T. Patterson, used under ,public domain',
licensing, cartography: Christian Sommer
(ROCEEH).
7 J.-M. Benito. Reproduced under licence
'public domain' (Author: Locutus Borg).
Accessed 14.09.2020.
8 M. Arzarello. Reproduction under licensing ,public domain'. Accessed 14.09.2020.

Liane Giemsch
**From Africa around the world:
the Acheulean**
1 A. Gonschior/Heidelberger Akademie der
Wissenschaften
2 B. Groscurth/AMF/ROCEEH
3 from: Beyene, Y. u. a. 2013 The characteristics and chronology of the earliest
Acheulean at Konso, Ethiopia. PNAS 110, 5,
2013, 1584–1591, Fig. 4.
4 J. Vogel/LVR.
5 Atelier Wildlife Art for Hessisches
Landesmuseum Darmstadt,
Photo: W. Fuhrmannek, HLMD.
6 Data: ROAD (ROCEEH), background map:
T. Patterson, used under 'public domain'
licensing, cartography: Christian Sommer
(ROCEEH).

Liane Giemsch
Early human use of fire
1 J. Neskora on Unsplash.
2 B. Groscurth/AMF/ROCEEH
3 Data: ROAD (ROCEEH), background map:
T. Patterson, used under 'public domain'
licensing, cartography: Christian Sommer
(ROCEEH).

Angela A. Bruch, Karen Hahn

Raw or roasted? How fire changed what's on the menu

1 M. Schmidt

2 Data: A. Bruch/K. Hahn, Illustration: B. Groscurth/AMF/ROCEEH

3 Nr. 1 *Ficus sur:* Günter Baumann, African plants – A Photo Guide. www.africanplants. senckenberg.de

Nr. 2 *Hoslundia opposita:* Claude Boucher Chisale, African plants – A Photo Guide. www.africanplants.senckenberg.de

Nr. 3: *Brachystelma barberiae:* Wikimedia: Curtis botanical magazine pl.5607

Nr. 4 *Plectranthus esculentus:* Paul Latham, African plants - A Photo Guide. www.africanplants.senckenberg.de

Nr. 5: *Sclerocarya birrea subsp. caffra:* Claude Boucher Chisale, African plants – A Photo Guide. www.africanplants. senckenberg.de

Nr. 6: *Phoenix reclinata:* Jos Stevens, African plants - A Photo Guide. www.africanplants. senckenberg.de

Info box Baobab:

Baobabtree: K. Hahn; Leaves: A. Lessmeister; Fruits: F. Schrenk

Miriam Noël Haidle

Takink a detour on the path to human thinking

1 R. Walter

2 M. N. Haidle

3–6 B. Groscurth/AMF/ROCEEH

7 M. N. Haidle

Christine Michel

Of rattles and puzzle boxes – social learning as the key to being human

1 pixabay.com

2 S. Michel

3 Graphic based on: Kobayashi, H., Kohshima, S. Unique morphology of the human eye. Nature 387, 767–768 (1997). https://doi.org/10.1038/42842

Photos: galago: pixabay.com; macaque: pixabay.com; human: E. Quednau/AMF; baboon: pixabay.com; gorilla: pixabay.com, orangutan: pixabay.de; chimpanzee: Gemeiner Schimpanse: Zeppelin https://piqs. de/fotos/111667.html; gibbon: pixabay.com

4 Design: E. Quednau/AMF, Graphic based on: Michel, C./Pauen, S./Hoehl, S. Schematic eye-gaze cues influence infants' object encoding dependent on their contrast polarity. Scientific Reports **7,** 7347 (2017). https://doi.org/10.1038/s41598-017-07445-9

5 E. Quednau/AMF

Oliver Schlaudt

Habitus: The cultural primer

1 Martin Kraft, CC BY-SA 3.0, https://commons.wikimedia.org/wiki/Ovis_canadensis#/media/File:MK00658_Badlands_Bighorn_Sheep.jpg)

2 Stout 2002, Fig. 5

3 Photo: M. S. Müller, K. Behrendt, Source: http://www2.braunschweig.de/lichtparcours2016/lp16/presse.html.media/350757/_MG_0668.jpg

Miriam Noël Haidle

Across the mountains, into the wide world. Evidence of human expansion

1 H. Jensen, Uni Tübingen

2–7 Data: ROAD (ROCEEH), background map: T. Patterson, used under 'public domain' licensing, cartography: Christian Sommer (ROCEEH).

Roman M. Wittig

Chimpanzee cultures – a search for clues

1 C. Girard-Buttoz, Tai Chimpanzee Project

2 Ch. Boesch/Wild Chimpanzee Foundation

3 L. Samuni, Taï Chimpanzee Project (a–c), D. Morgan & C. Sanz (d).

4 B. Groscurth/AMF/ROCEEH, according to Fig. 1 from Whiten et al. 1999.

5 T. Gruber

6 L. Luncz

Info box Nutcracking

R. Wittig

Page 134 A. Gonschior/Heidelberger Akademie der Wissenschaften

Madelaine Böhme

Tracing the steps to becoming human: research and scientific methods

1–3 L. Giemsch

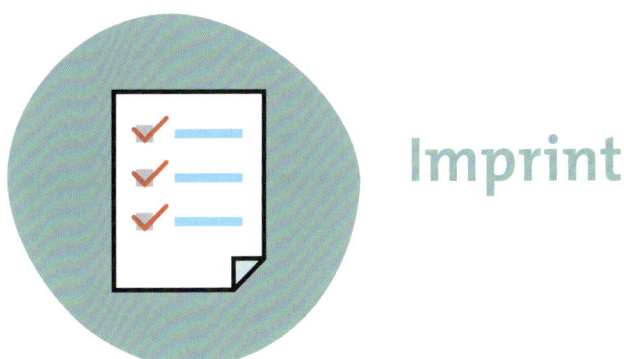

Imprint

Exhibition

Director Archäologisches Museum Frankfurt
Dr. Wolfgang David

Curators
Dr. Liane Giemsch, PD Dr. Miriam Noël Haidle

Scientific team
apl. Prof. Dr. Michael Bolus, PD Dr. Angela Bruch,
Dr. Liane Giemsch, PD Dr. Miriam Noël Haidle,
Dr. Karen Hahn, Dr. Christine Hertler, Dennis Hoffmann M.A.,
Dr. Christian Sommer, and the members of the student
project at Goethe University Frankfurt:
Anne Bachmann, Victoria Diefenbach, Annika Haas,
Anna Therese Hoffmann, Johannes Kratz, Sandra
Niggemann, Kirsten Scheinberger, Elaine Schneider,
Angela Senftleben, Kea Simonis, Marcel Wachtel

Exhibition coordination and in charge of lending
Dr. Liane Giemsch, Tessa Maletschek M.A.,
Dr. Andreas Sattler

Insurance
Tessa Maletschek M.A., Dr. Andreas Sattler

Exhibition architecture and design
Eike Quednau

**Graphic implementation
and design of the advertising materials**
Eike Quednau

Text editors
Kim Ottendorf M.A., Dr. Andreas Sattler

Technical services and exhibition setup
Wolfgang Block, Thomas Flügen,
Heinrich-Teja Huppertz, Lothar Kant

**Museum education and coordination
of the accompanying program**
Dr. Liane Giemsch, Maria Meßner M.A., Dr. Christian Peitz

Restoration support and exhibition facilities
Thomas Flügen, Sigrun Martins, Dipl.-Rest. Birgit Schwahn

Event management
Maria Meßner M.A.

Human resources and financial management
Nicole Fallert, Birgit Milanovic

Public relations, Press, and social media
Maria Meßner M.A., Sara Martin M.A.

Animation on human expansion
LVR-LandesMuseum Bonn, Architectura Virtualis GmbH

**Film on the production of a handax by Harm Paulsen,
experimental archeologist**
Camera: Claus-Peter Grätz
Thomas Claus Medienproduktion 2002
© Landesamt für Denkmalpflege und Archäologie
Sachsen-Anhalt

Multitouch applications
Programming and design
Bluelemon, Köln
Image research and acquisition
Dr. Tomáš Zachar

Maps in multitouch application and exhibit
apl. Prof. Dr. Michael Bolus, Annika Cüppers,
Dr. Christine Hertler, Dr. Andrew Kandel,
Dr. Christian Sommer

Diversity

Common
ancestors

Ardipithecus kadabba

Ar. ramidus

A. bahr-el-ghazali

A. afarensis

A. prometheus

Orrorin
tugenensis

Australopithecus anamensis

Kenyanthropus platyops

Sahelanthropus
tchadensis

Graecopithecus freybergi

7   6   5   4